SPIDERS

OF THE EASTERN UNITED STATES

A Photographic Guide

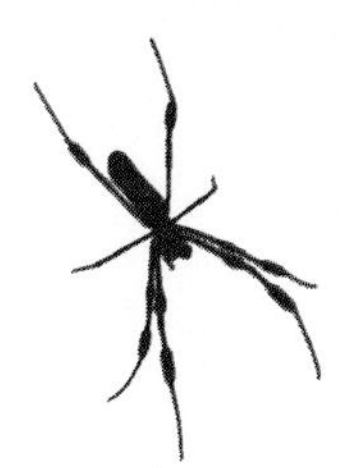

W. Mike Howell, Ph.D. and Ronald L. Jenkins, Ph.D.

Department of Biology, Samford University

To Greg and Spider –
I know that you guys love
spiders as much as I do!
W. Mike Howell

Cover photograph courtesy of W. Mike Howell.

Printed in the United States of America

10 9 8 7 6 5 4 3 2

ISBN 0-536-75853-0

BA 998586

JE

Please visit our web site at *www.pearsoned.com*

PEARSON EDUCATION
75 Arlington Street, Suite 300, Boston, MA 02116

Contents

I Acknowledgments

A compilation of information on the spiders of the eastern United States is an impossible task for two persons to do alone. Fortunately for us, we have had the unselfish help and encouragement of innumerable students, colleagues, friends and relatives. Knowing that we risk omitting the names of many persons who have contributed to our study, we must express our deepest appreciation to the following individuals: Mary Albert, Paul Aucoin, Mark Bell, Christine Boone, Reid Bryson, Tyler Burgess, Colin Chisholm, Maggie Clancy, Patrick Clancy, Stacie Clark, Diane Coffey, Connell Coffey, Dan Connell, Jennifer Connell, Brandy Covert, Larry Davenport, Betsy Dobbins, William Dobbins, Christopher Dodd, Andrew Dye, Adam Edwards, Jessica Evans, Dan Garrison, Stephanie Ash Gooch, Corey Guffey, Karen Hamilton, Drew Hattaway, Matt Hayes, Matt Henderson, Christina Hill, Mary Howell, Todd Howell, Holly Howell, Hunter Howell, Dawson Ingram, Allyson Ioos, Anna-Lea Jenkins, Ben Jenkins, Kitty Jenkins, Shannon Jordan, Joseph Kennedy, Rachel King, Tom Landry, Dave Langella, Karen Lee, Scott Linton, Kate McWhorter, Michael Measels, Jennifer Mitchell, Matthew Montz, Andrew Morrow, Virginia Morrow, Kevin Morse, Jennifer Musser, Jeff Newman, Joseph Park, Rose Parrino, Conner Patrick, Kimberly Patrick, Mike Patrick, Rorie Payne, Jason Rogers, Luke Roy, Phillip Royal, Chad Scroggins, Stefani Shadowens, Angela Sledge, Page Stanley, Jimmy Stiles, Robert Stiles, Becky Strickland, Rebecca Thomas, Linda Wood, Dan Voelzke, Janet White, Mike Widra and Marcia Young. Ms. Heather McNatt is due a special thanks for her continuous interest in our book in its early stages and in her untiring collection of spiders from the southeastern U.S. Carl L. Ponder was relentless in his collections of spiders from the Tennessee River Valley of North Alabama and has our hearty thanks for sharing his field experiences, extensive knowledge and specimens with us.

We owe an especial debt of gratitude to the following spider experts for sharing aspects of their extensive knowledge of spider biology with us: Professor Winston Baker, Alexander City, AL., Dr. Cole Benton, Jacksonville State University; Dr. G. B. Edwards, University of Florida; Drs. Debbie and George Folkerts, Auburn University; Dr. Howard Hunt, Atlanta Zoo; and Dr. Norman Platnick, American Museum of Natural History.

Finally, we are grateful to the following administrators of Samford University who have given us the time, funds and encouragement to complete this book: Thomas Corts, President; Brad Creed, Provost; Roderick Davis, Dean, Howard College of Arts and Sciences.

J. William Mathews, Jr., V. P. Business Affairs and General Counsel is due our sincerest thanks for his tireless efforts in encouraging us and in helping to get this book published.

II
Introduction

A. THE HUMAN NEED TO IDENTIFY SPIDERS

The spiders of the eastern United States are poorly known by both scientists and laypersons alike. Therefore, the need for a book dealing specifically with spiders inhabiting this portion of the country has existed for some time. At one time or another, almost every person has had a close encounter with a unique spider that they have wished to identify. Perhaps most have only wanted to know whether it was a species venomous to humans or a beneficial one that preyed upon insect pests. Others have had a more academic interest in spiders, wanting to identify the creatures and learn something about their life history and habits. Until now, these people have had either incomplete, general references available to them or very complex scientific treatises. Therefore, it was our desire to produce a photographic guide to the identification of commonly encountered eastern spiders to fill this educational void. A geographic region the size of the eastern U.S., coupled with its complex topography and temperate to subtropical climate, give this region one of the richest spider faunas in North America. No one really knows just how many species of spiders inhabit this region. A good estimate would probably put the number between 800 and 2,000 species. To further complicate the problems with spider identification is the fact that spider classification among many spider families is poorly known. Plus, there are many spider species within the eastern U.S. that have never been formally named, and many more which await discovery by araneologists (scientists who study spiders). In spite of this difficulty, the average citizen is still fascinated by spiders and wishes to put a name on each kind that they encounter.

B. SPIDER IDENTIFICATION IS NOT EASY

The correct identification of a spider to the species level is difficult even for the trained spider specialist. Some reasons that spiders are so hard to identify are: (1) spider anatomy is complex and most of the anatomical parts important in species identification are small and difficult to see without a microscope; (2) of the few taxonomic keys available, most are out-of-date because of recent descriptions of new species and continual revisions of families; (3) taxonomic keys are based upon adult spiders and often will not work on juvenile or sub-adult specimens; (4) many sibling species (biologically different species which look similar) occur among some spider groups; and (5) the paucity of books on spider identification that are available, most of which are only general guides containing ink drawings of the most common of United States spiders.

C. THIS BOOK IS ONLY A PHOTOGRAPHIC GUIDE, NOT A SCIENTIFIC TREATISE

It is appropriate here to describe what our book *is*, and what our book *is not*. It is not a taxonomic key to eastern spiders nor a scientific treatise on eastern spiders. **It is a photographic guide to some of the most commonly encountered eastern U.S. spiders.** We have also included a few spiders that are common in nature, but are not often seen because of their cryptic habitats or habits (for example, the trapdoor spiders).

D. A WORD ABOUT OUR PHOTOGRAPHIC TECHNIQUES

All of our spider photographs and images are original and previously unpublished. All were done by the authors. Some photographs were made using conventional color film (ASA 200), a Nikon N90S camera equipped with 52.5 mm extension tube, a 105 mm AF Micro Nikkor lens, and a Nikon SB-23 flash. Manual camera settings were always used. All digital images were made using a Nikon D1 camera equipped with a 52.5 mm extension tube, an AF Micro Nikkor 105 mm 1:2.8 D lens and a Nikon speedlight SB-28DX flash. We built a simple, home-made flash-bracket that supported the flash over the front of the lens. It was angled down directly at the spider (Figure 1). The flash allowed us a great depth of field (*f*/32 or *f/22*), adequate lighting, and permitted us to shoot the spider even if it was moving. For those interested in spider photography, we suggest a book that we utilized heavily while making the photographs for our book. The book is entitled "*Photographing Butterflies and other Insects*", by Paul Hicks, published by Fountain Press Limited, Fountain House, 2 Gladstone Road, Kingston-upon-Thames, Surrey KT1 3HD, United Kingdom, 1997, 96 pages. Our camera set-up, along with instructions on how to build the home-made flash bracket, are described in Hicks' book. Actually Hicks' camera set-up is a facsimile of John Shaw's "butterfly bracket" set-up and is illustrated and described in detail on page 82 of Hicks' book. We highly recommend this camera set-up as excellent and almost fool proof for obtaining both quality photographs and digital images of small creatures such as spiders.

In our photographs and digital images, we tried to capture those aspects of the spider (color patterns, eye positions, relative leg sizes, etc) that were important in its identification. Our aim was to produce a sharp photograph/image so that most species could be identified by comparing an unknown spider (using a hand lens or magnifying glass) with our photographs/images. This, coupled with the use of other sources on spider identification (taxonomic keys, species lists, etc., listed in the bibliography near the end of this book) could allow a more positive identification.

FIGURE 1. Our photographic unit for spider photography based upon John Shaw's "Butterfly Bracket" set-up

E. HOW OUR BOOK CAME INTO BEING

Originally, this book grew out of our desire to produce a photographic guide that could be used by students in our field courses at Samford University as they tried to identify unknown spiders. It is now our sincerest wish that not just our students, but persons from all walks of life who are interested in spider identification and biology, may find our book of value. We realize that this book should not be used as a final resource for spider identification, but as a springboard to the more scientifically accurate and detailed spider literature as one seeks to learn more about the identification and biology of these fascinating creatures.

III
Spiders vs. Insects

A. A SPIDER IS NOT AN INSECT

Spiders belong to the Class Arachnida which also includes the scorpions, pseudoscorpions, whip scorpions, ticks, mites, daddy longlegs (harvestmen), and some other minor groups. Spiders are distinguished from these other arachnid relatives by being placed into the Order Araneae which includes some 35,000 spider species distributed worldwide. All spiders have a body consisting of two main parts, a **cephalothorax** and an **abdomen** (Figure 2). In spiders, the cephalothorax usually has a pair of chelicerae with fangs, a pair of pedipalps, and **four pairs of walking legs**. Antennae and mandibles (jaws) are lacking. Usually 8 eyes are present on the cephalothorax and these are simple eyes (ocelli) with lens, rods and a retina. Spiders never have wings. Spinnerets are always present at the posterior end of the abdomen and produce silk.

B. AN INSECT IS NOT A SPIDER

Insects belong to the Class Insecta which contains over 900,000 species. Insects are readily distinguished from spiders by having a body consisting of three main parts: **head, thorax and abdomen**. The head bears neither chelicerae nor pedipalps, but does possess mandibles and antennae. While the head of most insects bears three ocelli, these do not form images. The insect uses compound eyes for its vision. The compound eye may consist of literally thousands of individual visual units called ommatidia, thus differing significantly from the simple spider eye. The insect thorax usually bears **three pairs of walking legs** and two pairs of wings. Silk is normally not produced by most insects, but some insect larvae (e.g. silkworms) do produce silk from an opening on the lower lip (as opposed to the abdomen). Embioptera produce silk from glands on the forelegs.

C. SPIDER VENOMS VS. INSECT VENOMS

Spiders have internal venom glands in the cephalothorax and these are associated with hollow, hypodermic needle-like fangs on the chelicerae in front of the mouth. Insects, if they possess venom glands, have them associated with a stinger-like structure and open at the posterior end of the abdomen. All spiders are predators, usually on insects. They seize their prey with their chelicerae and inject venom. Then the spider injects powerful digestive enzymes into and upon the prey, liquefying its tissues. It then simply sucks up the digested broth of its prey into the stomach. Thus, digestion in spiders takes place before the prey is swallowed in contrast to the insects which usually digest their food after they have swallowed it.

IV
Spider Anatomy

A. THE TWO BODY REGIONS

The spider body has two distinctive parts: an anterior **cephalothorax** and a posterior **abdomen**. A narrow **pedicel** connects the two regions. The cephalothorax bears the **eyes, mouthparts (a pair of chelicerae** bearing a **basal portion** and the distal **fangs**, and **pedipalps)**, and **legs**. The abdomen bears the openings of the reproductive, respiratory and digestive systems, and the **spinnerets** through which silk strands are pulled from the body (Figure 2).

B. THE CEPHALOTHORAX

The cephalothorax is covered dorsally with a shieldlike plate called the **carapace**. The anterior portion of the carapace is the cephalic or **head** portion that may be slightly separated from the posterior portion or **thorax** by the **cervical groove.** Immediately behind the cervical groove in many spiders is a **dorsal or thoracic furrow**. Three pairs of **radial furrows** may be present behind the thoracic furrow. Ventrally, the cephalothorax is covered by a median anterior plate, the **labium**, an anterior pair of basal segments of the pedipalps, the **endites**, a larger median posterior plate, the **sternum**, and four pairs of basal segments of the legs, the **coxae** (Figure 2).

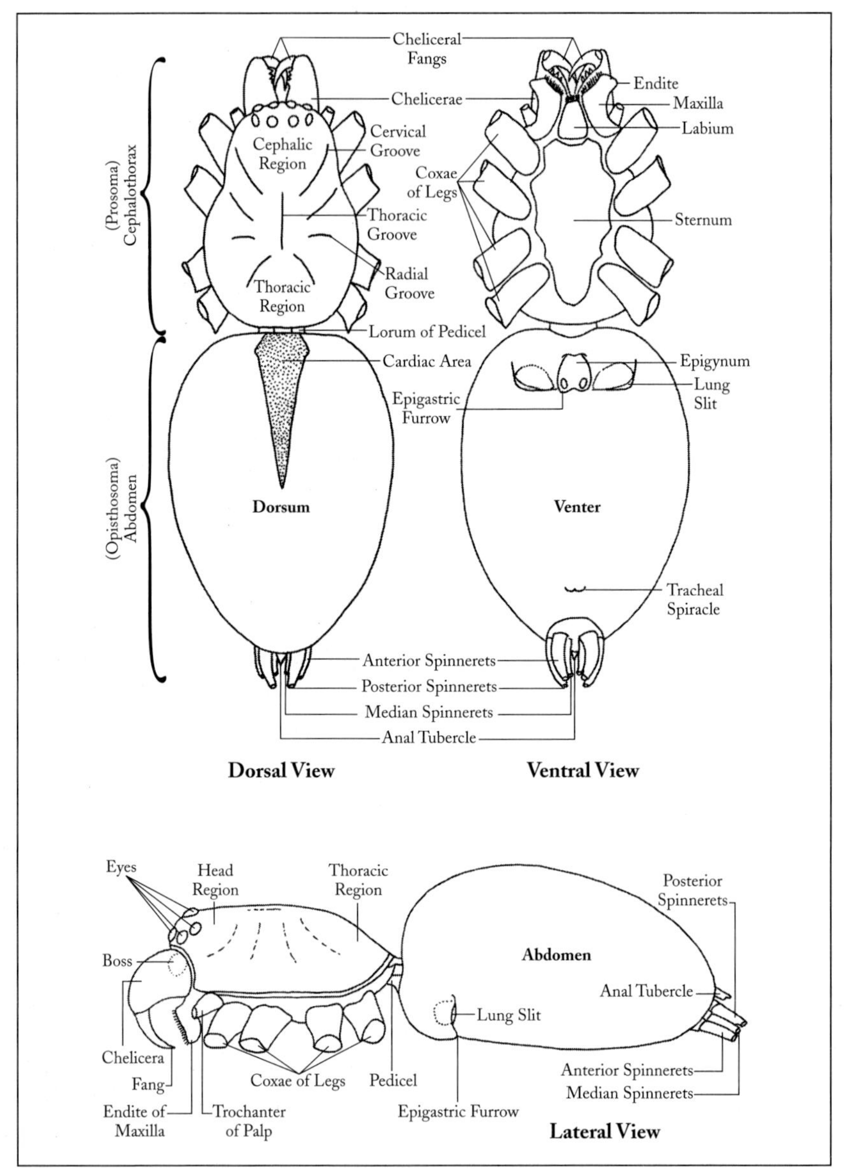

FIGURE 2. External anatomy of a generalized spider showing dorsal, ventral and lateral views (legs and palps have been removed). Drawing modified after Kaston (1978).

C. THE CHELICERAE AND FANGS

The first pair of head appendages is called the **chelicerae** or jaws, each of which is composed of a **basal part** and a distal **fang**. The fang is hinged or articulated to the basal portion. The chelicerae are located just below and in front of the clypeus. The basal part of the chelicerae often has a bump or swelling, called a **boss**, along the outer edge near the clypeus. A duct from the venom gland passes throughout the length of each chelicera and opens near the tip of the fang. In the primitive suborder of mygalomorph spiders, the chelicerae project forward and the fangs are articulated so that they move parallel to the median plane of the body. In the more advanced araneomorph spiders, the chelicerae extend downward, or obliquely forward, with the fangs articulated so that their movements are restricted within a transverse plane (Figure 3). The fang, when not in use, lies within the **cheliceral fang groove**. This groove may be provided with, or without, **promarginal teeth** on the anterior or upper side, and **retromarginal teeth** on the posterior or lower side. The number and position of the teeth on the cheliceral groove are important in spider identification and classification. Some spider groups have **stridulating organs**, each composed of a horizontal row of small grooves forming a **file** on the lateral surface of the chelicera, and a small projection called the **pick** on the pedipalp. By rubbing the pick on the file, they may make a sound. The mouth is the opening into the digestive tract and is located just beneath and behind the chelicerae (Figures 2–4).

D. THE PEDIPALPS

The second pair of head appendages, the **pedipalps**, are located just behind the mouth. Like the chelicerae, they are also considered mouthparts. Each pedipalp resembles a leg, but consists of six segments, rather than seven. The basal pedipalp segment is expanded on each side to form the **endites**, or **maxillae**, which are used to crush or manipulate

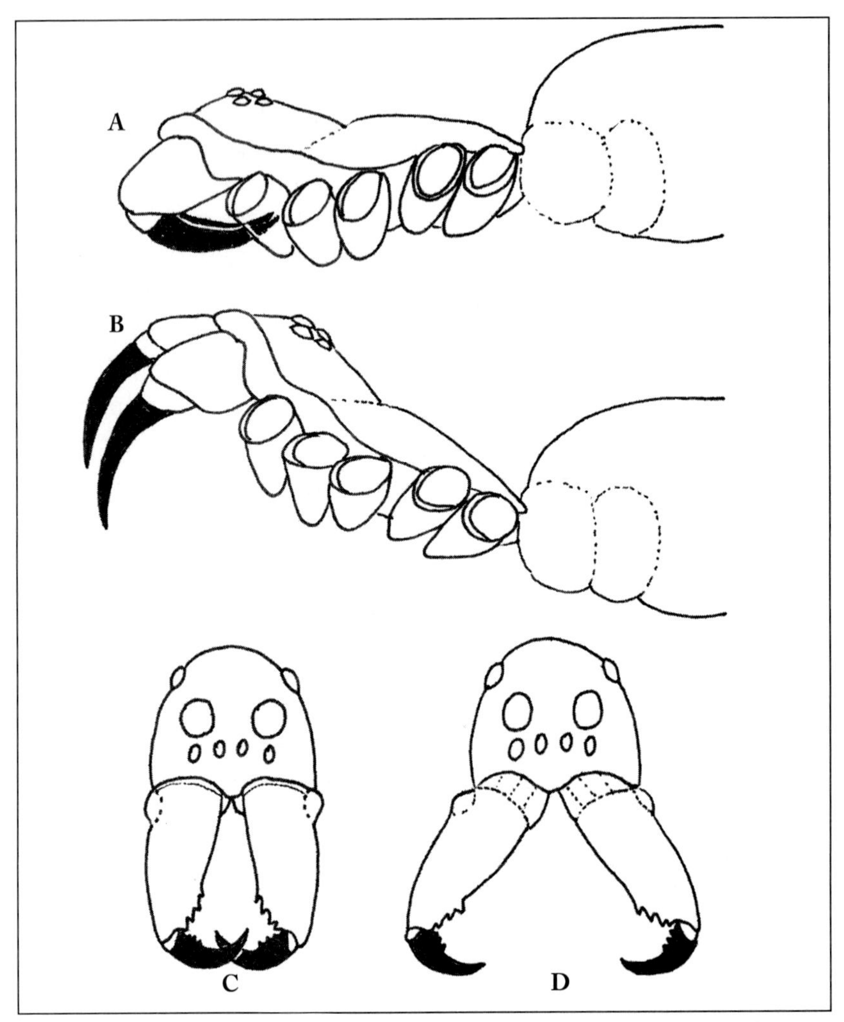

FIGURE 3 A-D. A comparison of the fangs of primitive spiders versus those of the more advanced spiders. (**A-B**) fangs typical of primitive spiders such as trapdoor and purseweb spiders (mygalomorphae). (**A**) fangs in resting position; (**B**) fangs and cephalothorax raised vertically in defensive and striking position; (**C-D**) fangs typical of the more advanced spiders such as orbweavers and wolf spiders (araneomorphae): (**C**) fangs in resting position; (**D**) fangs open in defensive and striking position. Drawing modified after Kaston (1978).

prey. The remaining five segments of each pedipalp (or simply **palp**) consist of the **trochanter, femur, patella, tibia and tarsus.** The palp also differs from the legs in lacking the *metatarsus,* the segment that, in the legs, occurs between the tibia and tarsus (Figure 4).

One of the most important functions of the pedipalps is in mating. Female spiders have unmodified pedipalps that look like walking legs. Penultimate or mature male spiders have an enlarged tip on the pedipalp, the **palpal tarsus** (or **palpal organ**), which is modified for containing a complex copulatory organ. Thus, the palpal organ allows for the easiest and quickest method for identifying the sex of a spider (Figure 5). Often, the male palpal organ is modified into a bowl-shaped structure called the **cymbium**. In some web building spiders, mature males are also provided with an appendage, the **paracymbium,**

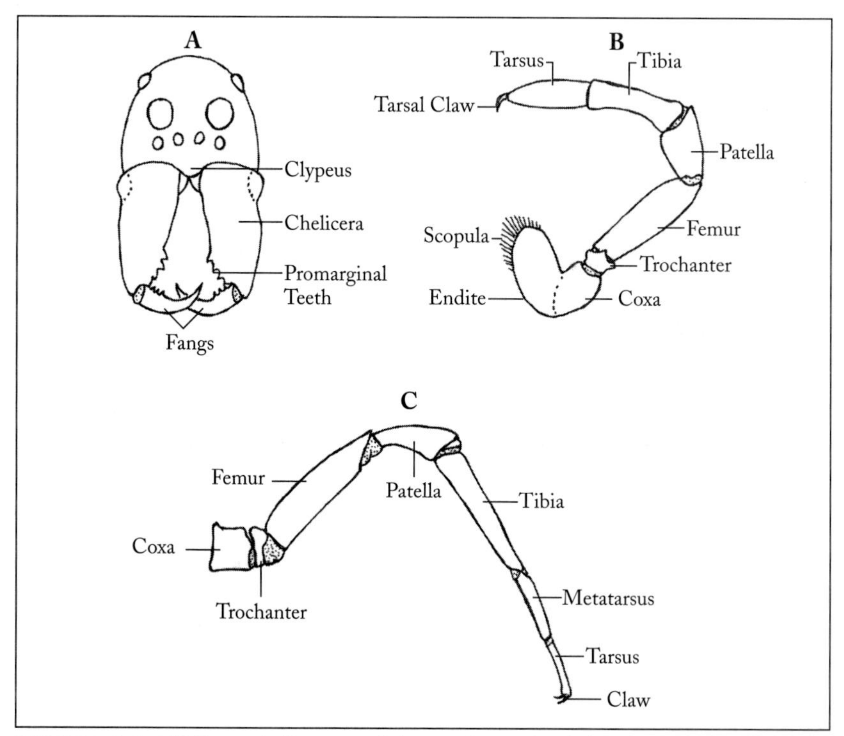

FIGURE 4 A-C. (**A**) Face, chelicerae and fangs of a spider; (**B**) pedipalp of a female spider; (**C**) leg of a spider. Drawing modified after Kaston (1978).

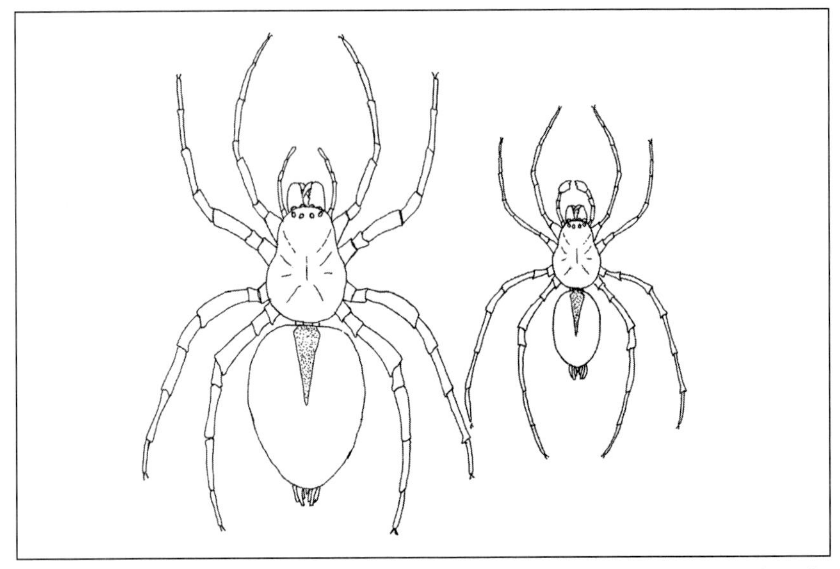

FIGURE 5. Comparative drawings of mature female and male spiders. Males (right) may be distinguished by their enlarged, modified tips of the pedipalps where sperm are stored. Adult females (left) lack the modified pedipalps. Males are usually smaller and have relatively longer legs than females. Drawing modified after Kaston (1978).

which originates at the base of the cymbium. The male palpal organ is inserted into the female epigynum to transfer sperm during mating. The complex structure of the male palpal organ is used by arachnologists to identify species and genera. For two reasons, we chose not to use the male palpal organ and the female epigynum to identify spiders. First, a dissecting microscope is needed to see these tiny structures. Second, much knowledge of the intricate, microscopic anatomy of the male palpal organ and the female epigynum is needed to identify closely-related spider species. Such knowledge is of little practical use to anyone except those professionally-trained spider taxonomists and systematists who must rely on these structures in order to determine species identity and relationships.

E. THE EYES

The anterior end of the carapace usually bears eight (four pairs) of simple **eyes** or ocelli. A few spider families have only six eyes and rarely, eyes may even be absent in some cave spider families. In most spider groups, the eyes are arranged into two rows. The position of the eyes in the two rows allows them to be designated in pairs as follows: **anterior medians (AME), anterior laterals (ALE), posterior medians (PME)** and **posterior laterals (PLE).** Six-eyed spiders lack the anterior medians. The two eye rows may be curved so drastically that it may appear to be three or four eye rows. An eye row is **procurved** when the lateral eyes are farther forward than the median eyes, or **recurved** when the median eyes are farther forward than the laterals. The space encompassed by the eye rows is the **eye area** or **ocular quadrangle**. The area enclosed by the four median eyes is called the **median ocular area.** The space between the anterior eye row and the front edge of the carapace is known as the **clypeus.**

The eyes are said to be **homogeneous** when they are all of the same light or dark color, or **heterogeneous** when they differ. Usually the anterior medians are dark in color and circular. The other eyes may also be dark and circular, or they may vary in color from clear to pearlescent or opalescent, and they may vary in shape from oval to triangular. Number, relative size, and spacing of eyes are of considerable value in arranging them into taxonomic groupings (Figure 6).

F. THE FOUR PAIRS OF LEGS

Spiders are equipped with four pairs of legs. Each leg is composed of seven segments. From proximal to the distal end of each leg, these segments are: **coxa, trochanter, femur, patella, tibia, metatarsus** and **tarsus** (Figure 4). The coxa, trochanter and patella are usually the smallest segments, with the femur being the largest. Roman numerals are often used to designate the legs. The most anterior leg pair is leg I, followed by leg

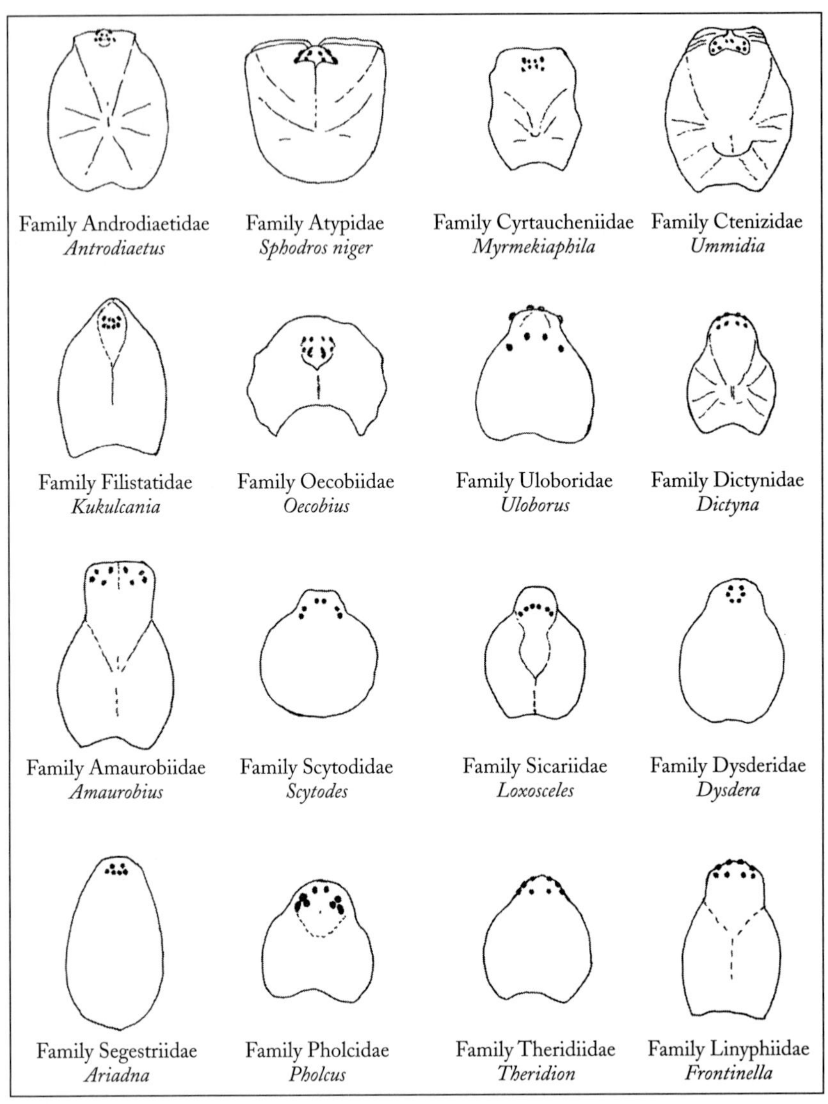

FIGURE 6. Dorsum of the cephalothorax showing number and position of eyes in spiders representing thirty-one different families. Eye number and position are important in identification of spiders.

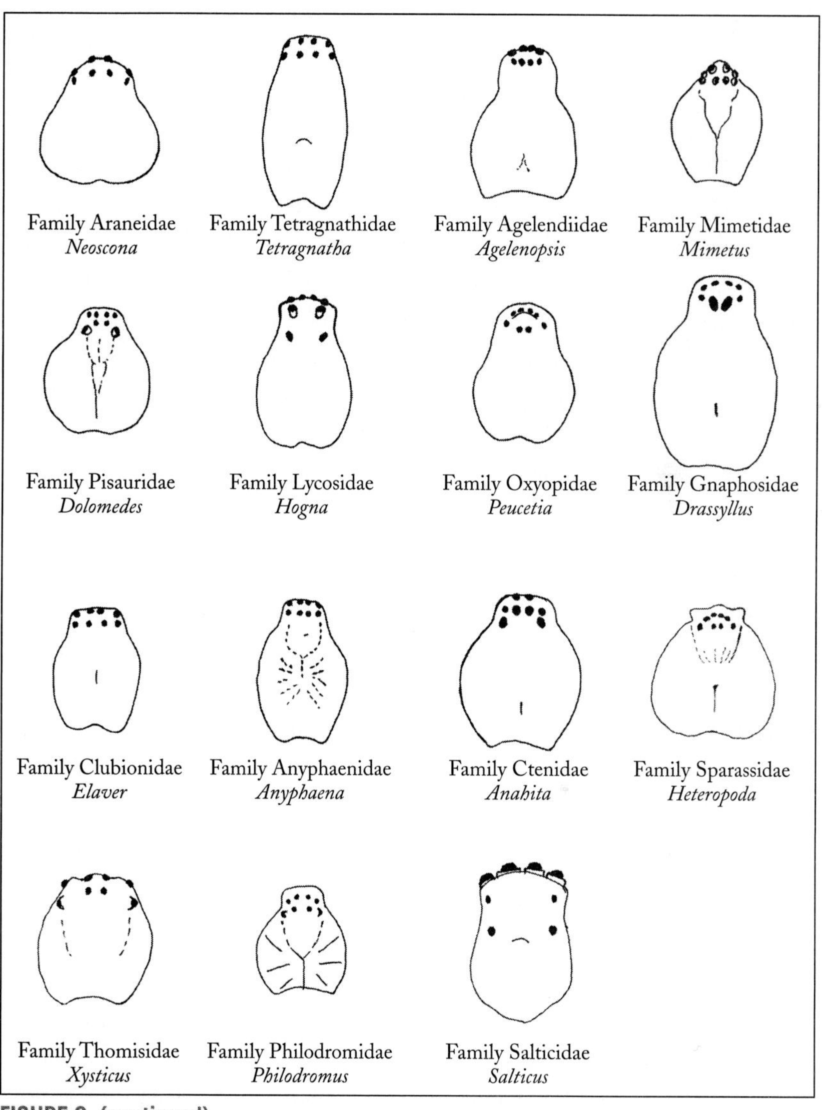

FIGURE 6. (continued)

II, leg III, and leg IV. The legs usually bear hairs, spines and scales. Often, very fine hairs, called **trichobothria**, are present. These are set vertically in tiny sockets and are important in identifying certain spider groups. The tarsus, or end segment of each leg, bears two or three claws and small hairs that may be clustered into **claw tufts** or pads. Often the ventral surface of the tarsus, and sometimes the metatarsus, is covered with a dense brush of short stiff hairs called the **scopula**. Theridiid spiders have a row of serrated bristles forming a **comb** along the ventral surface of tarsus IV. Cribellate spiders have, either on the dorsum or dorso-retrolateral edge of metatarsus IV, a large number of curved bristles forming the **calamistrum.** All of these leg structures are used in spider identification and classification.

G. THE PEDICEL

The slender, stalklike **pedicel** connects the cephalothorax to the abdomen. A sclerotized plate, or series of plates, called the **lorum**, covers the dorsal surface of the pedicel and the shape of these plates are of taxonomic value in separating lycosids from pisaurids (Figure 2).

H. THE ABDOMEN

The dorsal and ventral regions of the abdomen are seen in Figure 2. Abdominal shape varies considerably among and within spider groups and may be either globose, oval, elongate, triangular, or truncate. It may be smooth, or bear spines, humps or tubercles or have other projections. The abdomen is usually soft, but some spiders possess hard plates, called sclerites or **scuta**, which are variable in size, number and shape and have taxonomic value. On the anterior and middorsal part of the abdomen is the **cardiac area** (region above the heart). It is often marked with some distinctive pattern. Sometimes the whole dorsal abdomen has a pattern or **folium**.

On the ventral surface of the abdomen near the anterior end are a pair of transverse **lung slits** each of which opens into a **book lung**. The primitive mygalomorph spiders, such as tarantulas, possess two pairs of book lungs. Most families of spiders have an additional respiratory opening, the **spiracle,** which leads internally into respiratory tubes called **tracheae.** The single spiracle may be positioned either in the midline of the midventral region of the abdomen or at the posterior end just in front of the spinnerets. Joining the two book lungs is a groove, the **epigastric furrow.** Just anterior to the epigastric furrow are the reproductive pores. Female spiders usually have this area sclerotized and covered with a platelike structure, the **epigynum**, with a pair of slits used in mating to receive sperm packets from the palpal organ of the male. In some spiders, the epigynum has a finger like extension called the **scape**.

At the posterior end of the abdomen is the **anus** that usually opens on a small, cone-shaped **anal tubercle**. The **spinnerets** are located on the underside of the abdomen (venter) just in front of the anal tubercle. Most spiders have three pairs of spinnerets, but a few have only two pairs, while the most primitive family, the Liphistiidae in Asia, have four pairs. The median pair is usually the smallest and hidden by the anterior and posterior pairs. Silken threads are pulled from the spigots on the ends of the tubular or conical shaped spinnerets. Some families have a perforated plate-like structure in front of the anterior pair of spinnerets called the **cribellum**, from which a special kind of silk is formed. The cribellate spiders which possess such a structure are provided with a calamistrum (comb-like) on metatarsus IV to comb the multistranded cribellate silk fibers. Many of the spiders that lack a cribellum (ecribellates) are provided with a small, conical appendage, the **colulus**, which is situated between the bases of the anterior pair of spinnerets.

I. ANATOMICAL TERMS IN THE GLOSSARY

For many anatomical and other terms not covered in this introduction, the reader should consult the "Glossary" near the end of this book.

V
Spider Silk and Webs

A. THE MANY USES OF SPIDER SILK

The ability to produce silken threads is the most distinctive feature of spiders. The woven pattern of the snare or web, cocoon, and egg sac can be used to identify the family, genus, and in some cases, the species of a spider. Spider silk is chemically similar to the silk produced by caterpillars. However, spiders produce silk throughout their life, whereas caterpillars restrict the act to a larval stage.

Silk in spiders is produced by the silk glands located in the last two body segments (segments 10 & 11) of the abdomen. Each silk gland produces liquid silk which is drawn from a great number of spigots on the spinnerets by the hind legs. As it is stretched and exposed to the air, the liquid silk solidifies within seconds. Therefore it must be rapidly molded into either egg sacs, cocoons, drag lines or snares.

Among the many species of spiders there are six types of silk glands each producing silk of different qualities. One type of gland produces silk used only in the production of egg sacs and is found only among female spiders. Another type of gland produces silk of the fibrous nature which is used to make the strands of the snares and is common among the web building spiders. Another type of silk gland produces the silken glue used to bind together the strands of certain types of webs. Still another type of silk gland provides a sticky, glue-like substance which maintains the viscous coating of certain types of webs.

B. THE CHEMICAL NATURE OF SPIDER SILK

Chemical analysis of spider silk reveals that it is a mixture of many different components. The principal component of silk thread is the protein, **fibroin**. In freshly produced silk, the fibroin has a molecular weight of 20,000 to 30,000 daltons. When fibroin solidifies, ten or more fibroin molecules align in parallel fashion and bind into strands of molecular weights of 200,000 to 300,000. Some of the large fibroin strands are bound with other fibroin strands to form large accordion-like sheets of proteins. Stacks upon stacks of accordion sheets are laid upon each other in parallel fashion, running the length of the silk thread. Filling the empty spaces between the fibroin strands and protein sheets are loosely arranged amino acids. These amino acids contribute to the rubberlike properties of fresh silk. Another component of silk found in some spiders is large molecules of mucopolysaccharide which have hygroscopic properties, allowing excreted strands to take up water from the atmosphere. Certain mucopolysaccharides provide the stickiness for the strands of orb webs, while other mucopolysaccharides provide the glue that binds strands of the web together.

C. THE PHYSICAL NATURE OF SPIDER SILK

The physical properties of the silken threads of a spider web change as it ages. A strand of freshly excreted silk has greater elasticity and can be stretched three times its length. After several days of drying, silk threads will break if extended one-third their length. The tensile strength, or the weight that a silk thread can support, has been compared to many other natural products. The tensile strength of fresh silk exceeds that of a similarly sized tendon, bone material, cellulose, and rubber. However, a silk thread the same diameter as a steel thread, has only about half the strength of the steel (Vollrath, 1993). With age silk threads lose their elasticity and become more stiff, more brittle, and less sticky. These changes are brought about by changes in the hygroscopic properties of the mucopolysaccharides rather than the fibroin components.

D. OLD SILK IS RE-USED TO MAKE NEW SILK

Old silk is remarkably conserved by web builders. Peakall (1971) experimentally demonstrated the ability of an *Araneus* orb weaver to eat and re-utilize the fibroin protein of its old web. The old web was labeled with the radioactive amino acid, 3H-alanine. Thirty minutes after the spider was allowed to eat the old webbing, nearly 90% of the radioactive protein was re-excreted as new silk with rejuvenated elasticity and tensile strength. Because nearly all of the protein of the old silk was reassembled into new silk in such a short time suggests that either the fibroin protein was not completely digested before reassembly into silk or that silk production must invest nearly all of the protein synthesis activity of the spider.

E. SPINNERET NUMBERS, CRIBELLUM, AND COLULUS

Most modern-day spiders possess three pairs of spinnerets. These are the anterior, medial and posterior pairs of spinnerets located on the last two segments of the abdomen. On the terminal segment of each spinneret are numerous minute valves or spigots from which silk is excreted. As a spider excretes silk from the spinnerets, each spinneret can be moved independently of one another.

There are many modifications of the spinnerets among the different families of spiders. Most arachnologists believe that the ancestral spider probably possessed four pairs of spinnerets. This ancestral trait is still evident in one small group, the Mesothelae spiders. The Mygalomorph spiders (tarantulas and trap-door spiders) have lost the two anterior pairs of spinnerets and have retained the two posterior pairs of spinnerets. Among the Cribellate spiders three pairs of spinnerets have been retained while the fourth ancestral pair has been modified to form the cribellum. The cribellum is a plate that lies just anterior to the spinnerets and possesses hundreds of fine spigots for the purpose of excreting silk. The silk threads from the cribellum are exceptionally thin, having a diameter 1/100th the diameter of conventional silk threads. The fine silk from the cribellum is manipulated by the calamistrum, a special comb-like structure on the metatarsus of leg IV. As the fine cribellar silk is laid down in the web of these spiders, it takes on a "wooly" appearance. Prey are entangled in the wooly silk which is an important adaptation since the silk of cribellate spiders lacks the mucopolysaccharides which provide stickiness.

Among some of the ecribellate spiders (spiders lacking a cribellum) one anterior pair of spinnerets is retained as a vestigial structure, referred to as the colulus. The colulus is found among the families of Araneidae, Linyphiidae, Theridiidae and Thomisidae. The colulus does not contain the tubules that lead from a silk gland nor is it a functional spinneret.

F. THE MANY FORMS OF SPIDER WEBS

Webs are constructed by several families of spiders as a snare for the purpose of capturing prey. The contour of the web can be used to identify a spider's family and genus, and, in some cases, species.

1. Irregular webs

The webs of the spiders of the family Theridiidae are highly irregular and seemingly without design. However, some authors feel that theridiid webs share much similarity to the webs of the Linyphiidae. However, theridiid webs lack the horizontal domes. The theridiid web consists of very irregular and loose threads which are coated with droplets of "glue." A prey that flies into the web becomes stuck and entangled in the silk threads. As the prey struggles, the threads break and then swing to entangle more threads. The alerted spider quickly climbs down and throws additional silk over the victim before delivering its venom.

2. Funnel webs

Funnel webs are common to the spider of the family Agelenidae. Funnel webs are flat concave silken mats that possess a funnel retreat at one end. The agelenid spider takes refuge in the retreat, rushing out only when prey blunders into its web.

3. Sheet webs

Sheet web weavers construct horizontal, slightly convex, delicate silken dome webs that are supported above and below by vertical threads attached to the substrate. The spider always hangs beneath the webbed dome. As insects become entangled in the sticky, vertical strands they fall into the dome and are captured by the spider. These spiders are in the family Linyphiidae.

4. Tubular webs

The tubular web of the purse web spiders of the family Atypidae is thought to be a prototype of the most primitive web. These burrowing mygalomorph spiders extend a silken tube 1 cm in diameter from the opening of their burrow upward for 15 to 30 cm. Insects that crawl over this elongated purse web alert the spider living inside the silken tube. The spider crawls up through the tube and pierces the tube with its long chelicerae to secure and capture its prey. After the prey is pulled inside the tube, the spider quickly repairs the silken tube.

5. Orb webs

Orb webs are the circular webs which are most frequently seen in trees and shrubbery around houses and in woodlands. Orb webs in the vertical plane are built by the ecribellate spiders of the families Araneidae and certain members of the Tetragnathidae. Some tetragnathids build horizontal or slanted webs. Orb webs constructed in the horizontal plane are common to the cribellate spiders of the family Uloboridae.

The first threads that a spider lays down in the construction of an orb web are the frame threads which support the perimeter of the web and secure it to nearby supports such as the branches of bushes and trees. Radial threads are then fabricated by the spider and these extend from the frame threads to the central hub of the web. The frame and radial threads lack sticky mucopolysaccharides and are rigid threads that provide the skeleton for the web. The last type of threads to be laid down are the spiral threads which consist of sticky threads spiraled from the periphery of the orb inward toward the hub of the web. As a spiral thread intersects a radial thread it is glued in place with mucopolysaccharides.

There are many modifications of the conventional orb web. Some orb webs possess a free zone in which several radial threads and spiral threads are lacking in a sector of the web. Free zones are most common in the webs of those spiders that do not sit in the constructed web but reside in the safe retreat of a curled leaf off from the web. These spiders will extend a "signal" thread through the free zone from the central hub to their retreat. When the snare has captured a prey, the vibrations of the signal threads alerts the hidden spider. These spiders need not expose themselves to predation except when they are tending to their prey.

Among the argiope, or garden spiders, the orb web is supplied with a heavy zig-zag silken band that extend from the central hub toward the periphery between two adjacent radial threads. These heavy zig-zag bands form a structure called the stabilimentum. It has been hypothesized that the stabilimentum provides a number of possible functions, including stability to the web, camouflage for the spider, and as a warning to birds to avoid flying into the web. Because the stabilimenta reflect

UV light, it has been hypothesized to attract various pollinating insects into the web (Craig and Bernard, 1990).

Web building is a exceptional process among the orb weaving spiders. They have poor vision and rely almost totally upon their sense of touch and a peculiar sense of a 3-D orientation in the fabrication of a web. Experiments have demonstrated that when the eyes of a spider are covered with opaque paint, they are capable of building a perfectly normal web. Evidently vision is not a necessary requirement for web building. In the space capsule of Skylab in 1973, two argiope spiders were able to construct webs in complete weightlessness. However, the configuration of these webs could be disturbed if the webs were reoriented during the construction process (Witt, et al., 1977). Thus, orb web building spiders have a sense of orientation in space which is independent of gravity.

Most nature observers have noticed a fresh web in the early morning that was not there during dusk of the previous day. An ambitious orb weaver can lay down a complete web in only minutes during the evening hours. The frame and radial threads can be constructed in five minutes. The sticky spiral threads can be laid down in another thirty minutes. Because many orb weavers are nocturnal, fresh webs seem to appear each morning at sunrise. However, these may not last long as many orb weavers consume and remove their orb web soon after dawn breaks.

VI
Spider Ecology

A primary goal of this book has been to give a representation of the different species of spiders in the eastern United States by way of color photography. Equally important, we have sought to examine the interactions between spiders and the other members of the biological community. This, however, becomes a difficult task because we often think too small and simplistically in terms of the environment. We fail to realize the true complexity of the working assemblage of interacting organisms in the spider community. While there is a great diversity of spider species in the eastern states we must also look at the organisms with which they interact. It should be realized that spiders are a unique sector of a complex ecological "web."

ARACHNOPHOBIA AND ARACHNOPHILIA: NOT ALL SOCIETIES REACT THE SAME TO SPIDERS

Many societies around the world, including modern-day citizens of the U. S., have a fear of spiders (arachnophobia). Except for a couple of species that are venomous to humans, such fear of spiders is often unfounded and is usually related to our ignorance of these creatures. Strangely, a few societies do not actually fear spiders but have a reverence for them (arachnophilia). An example of an arachnophilic society would be the North American Indians. They were fascinated by the spider's ability to weave their webs as artists and as engineers. Native Americans envisioned a goddess, Spider Woman, capable of great creative spiritual forces. At the dawn of creation, Spider Woman wove and permanently bound together the distant east with the distant west. Then as creationist, she created human beings from earthen clay. There is still a small part of this respect and wonderment of spiders in the 20th century American society. The cartoon character Spiderman is a noble hero that accomplishes superhuman feats and with superhuman morality. An excellent account of arachnophobia and arachnophilia in human societies is given in a fascinating book by Hillyard (1994).

SPIDERS AS HABITAT SPECIALISTS AND GENERALISTS

In the temperate forests of the eastern U. S., spiders occupy essentially every microhabitat and some species even populate freshwater habitats. There is an entire spectrum of habitat specificity among the spiders. Some species are highly restricted in their preferred habitat, such as the crab spider, *Misumena vatia,* which frequents the blooms of the golden rod or milkweeds in silent stealth of flying prey that come to feed on nectar. The green lynx spider is equally specialized to vegetation and flowers and only reluctantly gives up its particular stem or bush. *Arctosa sanctaerosae,* a white wolf spider, is confined to the sands of the white beaches of the northern rim of the Gulf of Mexico. *A. sanctaerosae* is so well camouflaged against the beach sands that it almost never ventures into the dune vegetation. Tetragnaths are less restrictive in that they construct orb webs in various types of vegetation along streams, creeks, swamps and lakes.

A good example of a habitat generalist would be *Phidippus audax,* a common species of jumping spider of the Family Salticidae. This spider stalks its prey during daylight hours in a vast array of microhabitats, from secluded forests to "suburbia USA." The common house spider, *Achaearanea tepidariorum,* of the family Theridiidae, is such a generalist that it will build its irregular web in any dry, secluded site: beneath rocks, high in shrubs and trees, in caves or rock overhangs, beneath stumps, or the darkened corners of our homes. Habitat generalists are also seen among the many types of wolf spiders, such as *Hogna carolinensis,* which are active hunters prowling the surface of the ground in search of prey in all type of habitat from forests to meadows, and even our own backyards.

SPIDERS AND THEIR FOOD

Spiders are ubiquitous predators in every terrestrial ecosystem. Most spiders are not particular about what they eat, feeding on most any type of insect within their reach. However, spiders do consume other non-insect arthropods and even other spiders. The fishing spiders of the genus *Dolomedes* prefer small fish even though they will accept small aquatic invertebrates.

The victims upon which a spider preys are usually smaller than or equal to the size of the spider. However, it is not out of the ordinary that a jumping spider or orb weaving spider will capture prey several times larger than themselves.

THE PRIMARY GOAL OF A SPIDER IS TO EAT

The relationship between the predatory spider and its prey is a product of millions of years of evolution. In this slow process, natural selection has favored the spider that is less particular about the species of prey that it eats. This generalized feeding behavior or polyphagia has many advantages to the spider, namely that it assures that a spider will consume a greater dietary mixture and that its species stands a greater chance of survival when food resources become scarce. Riechert and Harp (1987) reared wolf spiders of the genus *Pardosa* in the laboratory and demonstrated that they were healthier spiders when fed a diversity of prey species than were the same species of spiders that received a restricted diet in the laboratory, or the spiders that survived in nature.

WELL-ADAPTED SPIDERS CAN "WAIT OUT THE FAMINE"

Natural selection has favored the survival of the spider that can "wait out the famine" and then rebound when food becomes available. We have attempted to house several species of trapdoor spiders including *Ummidia audouini* and *Cyclocosmia truncata.* These two species could never be enticed to eat. The *Ummidia* survived 187 days and *Cyclocosmia* slightly longer. Anderson (1974) starved a wolf spider, *Lycosa lenta,* for 208 days before it died. Web builders have lived in the laboratory for 276 days without eating. The ability of spiders to survive long-term starvation can be attributed to their low basal metabolic rate (BMR) and their sit-and-wait strategy in capturing prey. When it is feeding on a regular basis, a spider's BMR and level of activity increases. During periods of starvation both BMR and activity significantly decline.

WHEN FOOD BECOMES SCARCE, SPIDERS MOVE

When food becomes scarce at one location, spiders are inclined to move to sites with more available prey. Natural studies on the filmy dome spider, *Neriene radiata,* have demonstrated that these spiders lived in a given web for an average of ten days. When the diet of the spider was supplemented by the investigator, the homestead was occupied significantly longer for an average of 25 days (Martyniuk, 1983). The funnel web spider, *Agelenopsis aperta,* builds its web in forest floor vegetation in areas where the prey are most abundant. Additionally, this spider selects web sites where the temperature is most favorable for the spider and its prey

(Riechert and Tracy, 1975). Vollrath (1988) studied the orb weaving spider, *Nephila clavipes,* of the Panamanian forest. The spiders that constructed their large webs at the edge of the forest had their snares more exposed and available for prey capture than those webs that were built to the interior of the forest. Likewise, the spiders at the periphery of the forest captured and consumed more prey and they exhibited greater growth rates. Riechert and Tracy (1975) demonstrated that web site construction for the orb weaving spiders is carefully balanced between the favorable element of prey abundance and isolation versus the destructive factor of wind damage.

In another study of the web building habits of *Nephila clavipes,* Rypstra (1981) correlated the availability of food and kleptoparasites to their tendency to relocate their webs. In the web of *N. clavipes* there are often smaller theridiid food thieves (kleptoparasites) of the genus *Argyrodes* who wait in the periphery of the large *Nephila* webs and take prey that the host normally would have eaten. The greater the number of *Argyrodes* in a host web, the greater the tendency for the host to relocate its web.

Spiders are found only wherever the conditions are right for their insect prey. If you notice a spider residing in your house or yard you can be assured that there are also insects around that are being eaten.

SPIDERS DON'T LIKE ENVIRONMENTAL EXTREMES, ESPECIALLY THE COLD

Extreme temperatures are the most extensively documented abiotic mortality factor for spiders. The severe winter of 1978–1979 in the eastern United States killed many species of overwintering spiders. The basilica spider, *Mecynogea lemniscata,* was a rare find in the spring and summer of 1979. In many cases, extremely cold temperatures indirectly affect spider population survival through its negative effects on vegetation and insect prey populations (Rypstra, 1986). We have witnessed temporary drought and high temperature conditions that decimated spider populations in the southeastern U. S. during the summers of 1998 and 2000.

Many other abiotic factors can affect the survival of spider species. Heavy rains are destructive to many web builders. The desert agelenid, *Agelenopsis aperta,* has been devastated by heavy rainfall (Reichert 1974). Changes in sunlight can upset various orb weavers. *Micrathena gracilis* will abandon its webs when its pine forest is opened by logging or clear cutting operations. It will rebuild its web in the preferred closed canopy (Hodge 1987).

SPIDERS HAVE MANY ENEMIES

The most devastating enemies of a spider are wasps, birds and other spiders. Some wasps are parasitic, such as ichneumonids that lay their eggs on the outside of the spiders but otherwise leave the spider free. Fincke et al. (1990) counted 25–30% rates of ectoparasitism among juveniles of *Nephila clavipes* from ichneumonid wasps. Rehnberg (1987) documented a surprising rate of predation from mud dauber wasps on spiders. From the nests of the dauber wasp 792 prey items were removed and all were spiders.

Some birds feed on spiders. The Carolina Wren is common throughout the eastern U.S. and it is known to feed almost exclusively on spiders. During winter months when food sources become scarce, overwintering spiders are one of the few remaining prey for birds to eat. During spring and summer months of temperate forests the impact of predation on spiders by birds is not significant. When birds were excluded from spruce stands with netting, the mortality of the overwintering crab spiders and linyphiid spiders were significantly decreased (Askenmo et. al., 1977). In this study the size of the spider affected the rate of predation. In spruce stands that received no protective netting, the larger spiders were more commonly eaten than smaller individuals. In a tropical setting, Rypstra (1984) excluded birds (along with primates and large dragonflies) from regions of the Peruvian forest with netting. This was shown to significantly decrease spider mortality.

Spiders and lizards both feed on insects. However, lizards may also feed on spiders. Some Bahamian islands are devoid of lizards, likewise their density of spiders are higher than neighboring islands with lizard populations (Schoener and Toft, 1983). These investigators performed experiments that determined that the negative relationship from lizards on spider density of these islands is due largely to direct predation from lizards. However, the impact of lizards on common insect prey was minimal (Spiller and Schoener, 1988;1990).

COMPETITION DETERMINES WHERE SPIDERS ARE FOUND

Spiders are no exception to the principle of competitive exclusion which is well documented by natural and experimental evidence. This principle holds that no two species can simultaneously occupy the same ecological niche and compete for the same limited resource of nature. If they do, competition for the limited resource will become severe enough that one species will be eliminated or excluded from that niche by the other more aggressive species. The highly adaptive spiders have been able to avoid interspecific competition from other spider species by modifying their activity either spatially or temporally or they may even alter their life cycle. Throughout evolution spiders have been marvelously adaptive making room for new species of more adaptive spiders in an increasingly complex ecological web.

Common in the eastern USA are two orb weaving spiders that seem to occupy similar niches, *Mangora sp.* and *Metepeira labyrinthea*. Uetz et. al. (1978) determined that in forests where *Metepeira labyrinthea* constructed their webs between 0.5 m and 2.0 m from the ground, *Mangora* spiders only marginally invaded their territory. These *Mangora* spiders spun their webs from the ground up to 0.75 m. In another forest where *Mangora* dominated, the webs of *Metepeira labyrinthea* had been totally excluded. Among spiders that occupy very similar ecological niches and should be competitive, they seem to "give and take."

In another study, Brown (1981) compared the web sites for three species of large orb weavers, *Argiope aurantia, Argiope trifasciata,* and *Araneus trifolium.* When any one of the three species dominated a forest their webs were more randomly distributed. When these three species populated the same region of a forest, they constructed their webs at different heights from the ground. *A. trifolium* webs were significantly higher, *A. aurantia* webs were lowest, and *A. trifasciata* webs were intermediate in height from the ground. It was concluded that web height, as a means to avoid competition, attributed to 80% of the differences in foraging pattern.

The species of wolf spiders of the genus *Pardosa* have avoided competition by a number of modifications. Greenstone (1980) demonstrated that *P. tuoba* preferred the scrub and prairie habitat. However, *P. ramulosa* preferred aquatic habitats including freshwater streams, marine beaches, or small ponds. Suwa (1986) demonstrated that species of *Pardosa* of Japan altered the time of their daily activity and life cycle phenologies to avoid competition.

One of the most impressive and comprehensive studies on the interspecific competition among spiders was done by Gertsch and Riechert (1976). They used sophisticated statistical modeling to categorize a desert community of 90 spider species into groups that frequently interact. They determined that in most cases the many different spider species were separated from each other by spatial differences or temporal differences whether daily or seasonal. An underlying assumption by these authors was that competition between the adaptive spider species molded the evolution of this desert community.

SOME SPIDERS DON'T COMPETE, THEY MERELY INTERFERE

In the book, "*Spiders in Ecological Webs,*" David Wise (1993) eloquently details some well known cases in which closely related species of spiders cohabit the same area, at the same time, under the same conditions, and eat the same type of food. He cites his previous work on *Mecynogea lemniscata,* the basilica spider, and *Metepeira labyrinthea*. Both species build webs in the same sort of vegetation, at the same height, with similar phenologies, and compete for the same prey. He cites observations in which the two spiders have constructed webs so close together that the webs were actually joined. When females of one of these species were removed from their webs and placed in the web of the other species, both species had similar success rates of ousting the competitor. Clearly, these two species do compete for food; however, they successfully cohabit without an effect on survival or fecundity on the other species.

MOST SPIDERS ARE TERRITORIAL

Spiders are notorious hermits, preferring distant isolation from other adult individuals of the species. Experimental evidence has demonstrated that most every spider is territorial. *Araneus marmoreus* is a brightly colored yellow and black orb weaver. When the density of an *A. marmoreus* population is increased, their webs are more uniformly distributed. When the *A. marmoreus* population is thinned out and minimized, the webs become more randomly distributed. Thus, the territory of *A. marmoreus* extends beyond the web and its boundary is maintained only when population density is high (Reichert, 1982). Similarly, Janusz (1977) established that two species of wolf spiders were collected in pitfall traps with a more uniform distribution when the densities were high rather than when the population densities were low.

SPIDERS KICK THEIR YOUNG OUT OF THE HOUSE

Every species of spider must be efficient at dispersing its immature offspring. Dispersing the young assures that the solitariness of an adult spider will be maintained and that intraspecific competition for prey will be minimized. Some spiders achieve dispersal through ballooning, in which the young spin silken threads that are emitted into the blowing wind until the lift is great enough to "parachute" the tiny spider some distance from the parental web. Some spiders are known to balloon for miles and at altitudes of thousands of meters. Other species may balloon only a meter or so.

CAN SPIDERS KEEP INSECT POPULATIONS UNDER CONTROL?

Spiders are widely distributed and they may occur in great numbers. Bristowe (1939) estimated that 11,000 individual spiders cohabit an average acre of woodland in England. In 1971 he estimated one location is Sussex to contain 2,000,000 spiders per acre. Quick calculations would reveal that such a spider density throughout England would consume the biomass of insect prey equal to the British human population. Assuming that 11,000 spiders is a more realistic estimate of spiders in a temperate woodland, this high density must exert a major affect on insect populations. In one study, wolf spiders were found to be the most prominent control of centipedes in a North American grassland (Van Hook, 1971). Other studies have shown that spiders may be the best controlling predatory influence of collembola (springtails) in a beech and oak forest. Spiders had 2.7 times the predatory force as centipedes on the collembola population (Butcher and Zabik, 1970). In a classic experiment by Clarke and Grant (1968) spiders were removed by sieving the litter from four plots of a beech-maple forest, each 13 m^2. When spiders were excluded by sieving the litter and fencing the plots, the populations of collembola and centipedes were significantly increased but other arthropods such as millipedes were unaffected. It was concluded that spiders had a major direct predatory influence upon collembola and centipedes but not millipedes.

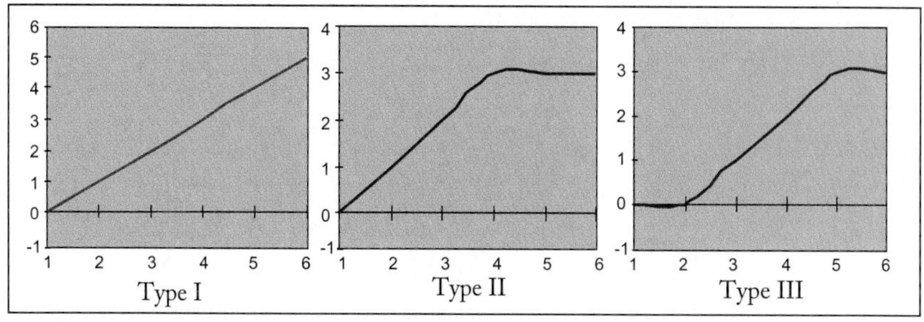

FIGURE 7. Type I, II, & III prey relationships, correlating prey density (x axis) and number of prey killed (y axis). Type I relationships represent the better prey control. Spiders are usually Type III.

Despite such examples as the wolf spider and collembola relationship, Wise (1993) convincingly argues that spiders in woodlands and grasslands may affect prey populations but they are not actually capable of regulating prey populations about an equilibrium. In ecological relationships in which the predator effectively controls prey populations, there is a linear relationship between the prey density and the rate of prey killed per time (type I). In type I relationships, there is an equal and effective control of the prey population regardless of the prey density. Most all species of spiders exhibit sigmoid relationship between prey density and the rate of prey kill per time (type III). At extremely low prey densities spiders capture few prey due to their sit and wait foraging behavior. At high prey densities, spiders become satiated and are not effective regulators of prey insect populations. Wise (1993) concluded that spiders may consume an impressive number of insects and they may aggregate toward higher prey densities and they may respond with greater fecundity, but they are not effective regulators of insect populations at low and high densities of prey. He concludes that the impact of spiders on insect populations is not tightly related to prey density in natural environments.

SPIDERS CAN BE USED TO CONTROL INSECTS IN CROP FIELDS

The role of some spiders can be more effective in insect control in agricultural settings than in natural woodlands and grasslands. Using ELISA techniques, Sunderland, et al (1987) examined the contents of the different predators in a cereal field for aphid proteins. Many species of predators tested positive for aphid consumption even when aphid densities were low. Spiders consumed 60% of the aphid population, greater than any other predator. One of the best experiments demonstrating the use of spiders in the agricontrol of insects comes from the study of Riechert and Bishop (1990) who designed a mixed vegetable garden that favored wandering spiders. They developed four different plots, each 70 m^2 that compared the effects of alternating rows of vegetables with rows of mulch and flowering buckwheat. The mulch created a habitat for wandering spiders and the buckwheat attracted pollinating insects. After the first and second years, the spider diversity was significantly increased by the mulch and buckwheat. This, in turn, significantly diminished the number of insect pests and the amount of crop damage.

In the agricultural environment, not all spiders are equally effective in the control of insects. Wandering spiders of the genera *Pardosa* or *Lycosa* have been effective in the control of agricultural insects. The lynx spiders, *Peucetia viridans* and *Oxyopes salticus,* have proven to be effective in consuming 85% of the arthropod pests and non-pests of cotton fields (Nyffeler et al.1987 a, b; Young and Lockley, 1985). Web builders have only negligible effects in the agricontrol of insects because their webs are easily destroyed (Nyffeler and Benz, 1981).

SPIDERS CAN HAVE A PROTECTIVE EFFECT ON PLANTS

Several studies demonstrate that spiders may protect certain plants by their consumption of insects which normally feed upon those plants. A simple experiment by Louda (1982) demonstrated a protective effect on the plants of a shrub community from the green lynx spider, *Peucetia viridis.* This spider tends to establish a territory in a given plant and captures prey as they come to feed on the flowers and foliage. On a single day at the end of the growing season, Louda selected twenty branches of the tallest flowering shrubs and compared ten that contained a green lynx spider with ten branches that did not contain the spider. The plants that contained a lynx spider had significantly greater biomass of foliage than the plants that did not have a lynx spider. However, the percentage of pollinated blooms in the flowering heads of the plants were significantly less than in the plants that lacked the lynx spider. The lynx spiders were active feeders on the insects that came to feed on the plant foliage such as leaf hoppers and aphids. It was also evident that the spider captured or detoured flying pollinators such as bees. The green lynx spider is a fine example that spiders can have a protective effect on plants but this positive impact also comes with a negative one, that of not being visited regularly by friendly insect pollinators.

HOW COMPLEX IS THE SPIDER COMMUNITY?

To fully understand the ecological impact of spiders it is important to consider every member of the community. It is near-sighted to think of the "spider community" as including only the spiders and their prey. It also includes the non-spider competitors, the vegetation upon which they live, and the abiotic factors that help to establish base-line living conditions. It is important to not restrict the "spider community" to merely its predator-prey relationships, but equate it to what many ecologist refer to as a "guild," the larger collection of interreacting organisms that cohabit a region. For example, the guild for wolf spiders would contain its competitors: scorpions, centipedes, and solpugid; its prey: ants, collembola, and other small arthropods; and its predators: scorpions, centipedes and other spiders. This intraguild relationship was compared by Polis and McCormick (1986) in a desert community following long term selective removal of either wolf spiders, scorpions, or solpugids from 300 fenced plots and a comparison of these experimental plots to 60 control plots. After 29 months, they uncovered various effects from the removal of scorpions, but not solpugids. When scorpions were removed from plots, the density of spiders significantly increased, while solpugids were unaffected. There were greater numbers of irregular and orb web constructed, as well as, greater numbers of wolf spiders collected in pitfall traps. Despite the removal of the scorpions, the number of collembola in the plots declined. Evidently, this was due to an increased predation pressure of spiders on collembola prey.

SPIDERS ARE A PARADOXICAL GROUP

In summarizing this chapter, it is evident that spiders are a peculiar group of animals which seem full of many paradoxes. Consider that spiders are designed to eat; however, their growth and reproduction rates are limited by food. Spiders are generally food limited; however, they rarely compete for food. Not being aggressive competitors, their ability to influence population densities of their prey is limited; however, they are well-adapted and ubiquitous predators in terrestrial environments.

These paradoxes have perplexed ecologists for many years and have been the focus of much disagreement among arachnologists. The bases of these apparent paradoxes is the non-competitive nature of spiders. How is it that a group of such diverse and successful animals have existed for an estimated 200 million years with only marginally competitive behavior? There are several possible answers. First, the sit and wait foraging pattern of many spiders that has proven so successful in their evolution requires them to be passive predators. Secondly, the density of a species of spider is usually so low in comparison to other guild members that they have only a marginal impact on prey densities. In unusual situations where prey densities explode and spider densities are elevated, the impact from the spiders can be appreciable. However, prey populations will be brought under control from guild members other than spiders. Another reason for the lack of competition from spiders is that spiders are diverse creatures which play almost as many roles in the environment as there are spider species. There is no one spider *persona.* Closely related species, such as the orb-weaver species of the genus *Argiope,* or the many species of wolf spiders, such as *Hogna* or *Rabidosa,* seem to share a very similar niche. However, upon closer inspection, we can paint a different ecological portrait for each species. It is not valid to pool the impact of all spiders into a singular role in the environment.

VII
Spider Classification

SCIENTIFIC NAMES

The biological world contains an incredibly complex variety of living organisms. In an attempt to categorize each kind of organism (species), scientists have developed an intricate system of classification and terminology. Spider classification, like that of all living organisms, is based upon natural relationships which are often inferred by studying an animal's morphology (structure) but which should eventually be shown to have a genetic basis. Our modern system of classification, or taxonomy, strictly follows an international code of zoological nomenclature (rules for naming organisms). Our present method of classification was developed by the Swedish naturalist, Carolus Linnaeus, in 1758, with some modifications since then. Linnaeus developed the **taxonomic heirarchy** whereby species were grouped into genera, genera into families, families into orders, and orders into classes. Today, we have expanded Linnaeus' hierarchic categories (taxa) to include, in an ascending series of increasing inclusiveness, the following taxonomic categories: *species, genus, family, order, class, phylum* and *kingdom.* Many biologists recognize that the only taxonomic unit that is real in nature is the **species**. All of the higher taxa (singular = taxon) are recognized as man-made units for conveniently grouping organisms which share similar characteristics. Even though it would seem to be a simple matter to recognize a "species" in nature, much modern-day disagreement and heated debate has focused on various "species concepts" that have been proposed. Despite the arguments, most zoologists adhere to the "biological species concept" which maintains that a species "is a group of interbreeding natural populations that is reproductively isolated from other such groups" (Mayr and Ashlock, 1991). This concept, while very practical to use, encounters difficulties when two or more populations are distantly separated in time and space. Then, the species status of the populations may be called into question. To overcome problems with this concept, a concept known as "the evolutionary species concept" has been proposed. It holds that "an evolutionary species is a single lineage of ancestor-descendant populations that maintains its identity from other such lineages and that has its own evolutionary tendencies and historical fate" (Wiley, 1981). However, the evolutionary species concept also has practical application problems which are beyond the scope of this book. We will, unless otherwise stated, follow the "biological species concept" as described by Mayr and Ashlock (1991).

The system we use to name species is known as **binomial nomenclature**. This essentially tells us that every species on earth will be given "two names" to constitute its scientific name. Each species is given a Latin (or Latinized) name consisting of two words (binomial). The first is the **genus** and the second is the **specific epithet**. The genus name plus the specific epithet, written together, constitute the scientific name of the species. The genus name is usually a noun, and the species epithet is usually an adjective that in some way describes the genus name. Thus, a scientific name often has meaning; for example, the scientific name for the "Starbellied Orbweaver" spider is *Acanthepeira stellata* (Marx). The genus name is derived from the Greek prefix, Akantha, meaning "thorn or spine" plus the suffix, epeiros, meaning "mainland or main body". The combination refers to the fact that the main body (or abdomen) of this spider has spines on it. The specific epithet, *stellata,* is an adjective derived from Latin and means "star-shaped; coming out in rays or points from a center". The scientist who originally described the spider as a new species and gave it its binomial name often has his/her name appended as a third part of the scientific name. In the example of the Starbellied Orbweaver, Marx was the scientist who first named it. We call this scientist the "author" of the species. Unlike the scientific name, the author's name is never italicized. The author's name appears after the scientific name either with or without parentheses around it. Parentheses indicate that the author, upon original description, placed the spider into a genus different from the one in which it is now placed. There are many reasons why the spider may have had its genus name changed. The most common ones are either the splitting or lumping of genera as new data on relationships become available and taxonomic revisions are published. In neither case, though, is the specific epithet changed. It should be noted that the term "species" is the same for both the singular or plural usage. There is no such term as "specie" in biology.

COMMON NAMES

In some well-known groups of organisms, common names are just about as useful, or more so, to naturalists and laypersons, than are the cumbersome-to-spell and hard-to-pronounce scientific names. Such a group where common names are presently more useful and recognizable than scientific names is the birds, whose common names have been standardized for several years. Unfortunately, there are so many species of spiders that many have never yet been given formal scientific names, much less common names. Fortunately, a committee was formed to begin the long laborious task of giving each species of spider a common name. The American Arachnological Society Committee on Common Names of Arachnids assembled, during 1995, a booklet entitled, "Common Names of Arachnids," published by the American Tarantula Society (Breene et al., 1995). While this is a good start, only a relatively few of our most common spiders were given common names by this committee. Many of the species in our photographic guide do not yet have a standardized common name, and for this we apologize to the reader. For those species that are given a common name, we have largely followed those common names as used in Breene et al.(1995), Fitch (1963), Jackman (1997;1999), Kaston (1978), and Levi et al. (1990).

VIII
Cribellate vs. Ecribellate Spiders

Early spider taxonomists considered the presence of a cribellum and calamistrum to be a fundamental phylogenetic distinction among the spiders of Suborder Araneomorphae (Labidognatha). Subsequently, the suborder was subdivided into Section Cribellate and Section Ecribellate, with the former section possessing a cribellum and calmistrum and the later section being more advanced and lacking them.

The cribellum is a sieve-like plate located anterior to the spinnerets and is thought to be modified from what originally was the most anterior pair of spinnerets. The cribellum possesses hundreds of small spigots from which the spider excretes broad ribbon-like fibers of silk. The cribellar silk is exceptionally fine stranded and is manipulated by combs of a series of bristles on metatarsus IV referred to a calamistrum. These strands forms the "hackled bands" spun into the web and facilitates the entangling of prey in the spider's snare. Common cribellate spiders in the eastern U.S. include the families Oecobiidae, Dinopidae, Filistatidae, Amaurobiidae, Uloboridae, and Dictynidae. All except the Dinopidae have been included in this book.

The majority of spider families are ecribellate spiders. They all lack a cribellum and calamistrum. Most produce various kinds of silk, some viscid and some non-viscid. But, none produce the hackled strands common to the cribellates.

Recent taxonomic work has not emphasized the importance of the division between cribellate and ecribellate spiders. Thus, in this guide we have adopted the sequence of families utilized by Platnick (1997) and have not segregated the families into these two sections.

IX
Descriptions of Families and Common Species

A. SUBORDER MYGALOMORPHAE

This suborder has also been referred to as the Orthognatha, a name which means "straight jaws", in reference to the parallel cheliceral fangs. Today, most workers use the suborder (or Infraorder) name Mygalomorphae which means "mouse body form", in reference to the fact that the group contains tarantulas and some other spider groups whose members possess large, hairy bodies with stout legs, somewhat reminiscent of a small mouse at first glance. The term Mygalomorphae, as used today, essentially means "tarantula body form." The large, tarantula-like purseweb and trapdoor spiders are in this group. Mygalomorphs have their venom glands located entirely within the chelicerae which are unusually large and powerful. There are two pairs of book lungs, the second pair present as white spots posterior to the epigastric furrow. Most mygalomorphs are tropical or subtropical, but a few species are common in the southern and western U.S. Raven (1985), in a classic work, extensively revised the mygalomorph families and some of the genera.

FAMILY ATYPIDAE—Purseweb Spiders

Purseweb spiders are large (up to 30 mm in length), tarantula-like mygalomorph spiders with abdominal tergites and six spinnerets. Unlike many other mygalomorphs, atypids are distinguished by possessing a transverse thoracic furrow, rather than a longitudinal or pit-like

FIGURE 8. Purseweb of *Sphodros rufipes* on base of dogwood tree.

depression. The sternum and labium are fused into one unit, without a separating seam or groove (Roth, 1993).

Atypids live within an enclosed silken-tube which extends from about four to eight inches underground up the sides of trees and rocks for a distance of up to ten inches depending upon the size of the spider. The diameter of the tube may be as large as two inches. Atypids in the U.S. usually construct a vertical tube attached to a tree, while those in Europe make tubes with the aerial portion horizontal to the ground.

The atypids are distributed in the eastern half of the U.S. There are two genera, and about eight described species. A widespread species in the eastern states, *Sphodros rufipes,* will represent these interesting spiders.

Family Atypidae—*Sphodros rufipes* (Latrielle)—Purseweb Spider

Identifying Characteristics: This large, dark brown to black spider has enormous chelicerae which are conspicuously projected forward. When viewed from above, the chelicerae are almost as long as the cephalothorax. Ventrally, the labium is fused to the sternum and this serves as a diagnostic feature of the family (Roth, 1993:22). They have six spinnerets, abdominal tergites, and a transverse thoracic depression. The coxae of each pedipalp is very well-developed, being elongated and tapered to form an endite, which serves as a mouthpart. The unique purseweb of silk that extends from beneath the ground up the side of a tree is also an identifying characteristic of this mygalomorph spider.

Ecology and Behavior: This spider builds a "purseweb" which is a tube of silk which extends from 4 to 6 inches below the ground to about 8 to 10 inches up the side of a tree. Above ground, the tube is covered with bits

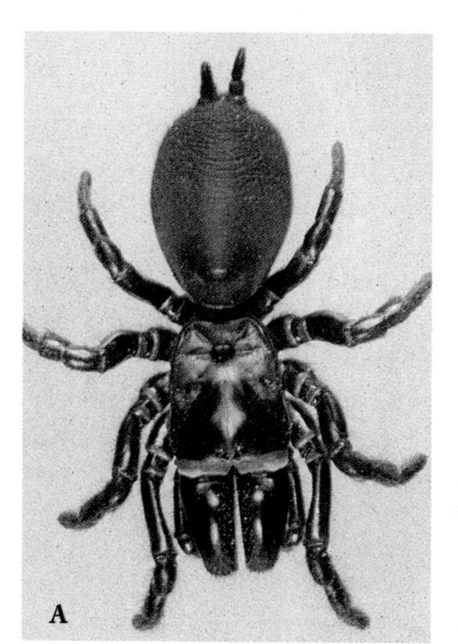

FIGURE 9 A-B. (**A**) Dorsal view, and (**B**) ventral view of large female *Sphodros rufipes*.

of leaves, lichens, algae and debris which form a very effective camouflage, helping the tube to blend in with the bark of the tree. Some specimens supposedly make tubular webs which do not extend vertically up the side of the tree, but rather, they construct a tube which extends horizontally over the ground and resembles a half-buried root. Even trained arachnologists often have difficulty in discovering a purseweb. The spider hides inside the tube until prey land on or walk across its surface. Then, the spider bites the prey through the tube with its enormous chelicerae. A slit is then cut in the silken tube and the prey is pulled inside. Afterwards, the undigestible remains of the prey are ejected through the slit and the spider repairs the tube and awaits another meal. This species is distributed from Maryland throughout the eastern U.S. to Texas. A similar species, *S. fitchi,* extends north to Kansas and Illinois (Gertsch and Platnick, 1980).

Size: Length of female 25 mm; of male 14.5 mm.

FAMILY ANTRODIAETIDAE—Folding Door Trap-Door Spiders

These mygalomorph spiders have abdominal tergites and a groove separating the labium and sternum. The females have a rastellum. The carapace has either a pit-like or longitudinal thoracic depression. The posterior lateral spinnerets have an inflexible, digitiform distal segment.

Antrodiaetids live in subterranean, silk-lined burrows capped with either a trapdoor or with a door consisting of two semicircular halves which meet in the midline to close the burrow's entrance.

Antrodiaetids are found throughout much of the U.S., but are particularly common in the southern tier of states from the Atlantic coast to the Pacific coast. This is a small family with three genera and 25 species (Roth, 1993; Coyle, 1968, 1971 and 1974). An antrodiaetid spider commonly found in the eastern U.S. is *Antrodiaetus unicolor.*

Family Antrodiaetidae—*Antrodiaetus unicolor* (Hentz)—Folding Door Trap-Door Spider

Identifying Characteristics: This large spider has a chestnut to cinnamon brown cephalothorax, chelicerae and legs. The chelicera extend from the head and are armored at the most anterior point with prominent rastella. A rastellum is a tuft of stiff bristles used in digging burrows. The abdomen is brownish gray in color and bears mid-dorsal tergites. Tergites are hardened plates that appear as raised dark spots or bands that are covered with numerous short bristles. The male possesses three dorsal tergites, the first and third being transverse bands and the middle being a circular spot. In the female the single tergite is less conspicuous than those of the male and is a homolog of the middle one in the male. Another distinctive feature of this trapdoor spider is the presence of four spinnerets, rather than the more conventional number of six. The anterior pair of spinnerets is short. However, the second pair is exceptionally long and extends beyond the abdomen.

FIGURE 10. A female *Antrodiaetus unicolor.*

Ecology and Behavior: In the eastern U.S., this large mygalomorph is less frequently encountered than some other species of trap-door spiders. We located the unbranched burrow of this spider in soft, sandy loam soil. This burrow was in a thicket of shrubs and small trees in a low-lying floodplain of a nearby creek. We have found this burrow and spider on the edge of cliffs in soil and moss in the rocky crevices. The burrow was thickly lined with silk webbing and extended to a depth of about six inches. The burrow often has a silken trap-door that folds in the midline. Sometimes the trapdoor is covered with moss or leaf litter. These spiders leave their burrows in the evening to hunt or repair their trap-door. During rainy seasons, the males may wander at night to escape flooded burrows. The range of *A. unicolor* extends from New York south to Georgia and Alabama and west to Arkansas (Kaston, 1978).

Size: Length of female 20 mm; of male 17 mm (Kaston, 1978).

FAMILY CYRTAUCHENIIDAE—Cyrtaucheniid Spiders

Cyrtaucheniid spiders are mygalomorph spiders that lack abdominal tergites and claw tufts. They are rather large tarantula-like spiders that reach 25 mm in body length. They are distinguished from the closely-related Ctenizid mygalomorphs by a combination of obscure anatomical features (Roth, 1993). These mygalomorphs live in subterranean, silken-lined burrows which are usually provided with trap doors.

These spiders are widely distributed throughout the U.S., but are especially common in the southwestern states. There are six genera, with about 15 described species and nearly twice that number awaiting scientific description. A common representative of the cyrtaucheniids that occurs in in the eastern U.S. is *Myrmekiaphila fluviatilis.* Based upon genitalia, this species is likely the composite of several species. The taxonomy of this genus is currently being worked out by Dr. Norman Platnick of the American Museum of Natural History.

Family Cyrtaucheniidae—*Myrmekiaphila fluviatilis* (Hentz)

Identifying Characteristics: A mygalomorph spider belonging to this genus can be recognized by having the cephalic portion raised above the thoracic portion of the cephalothorax. The chelicerae and cephalothorax are dark brown and the legs are lighter in color, often described as "cinnamon." The elongated abdomen varies from light to dark gray and it may possess abdominal markings. When abdominal markings are present there will be three, four or five darker transverse bands. We have collected one specimen that lacked the transverse bands but did possess a broad dark longitudinal band extending the length of the abdomen. Positive identification of a specimen of *Myrmekiaphila* must include the presence of a rastellum (a squared medial edge of the chelicera that the spider uses to dig its burrow). Additionally, the ventral surface of the sternum will have two sigilla, small hairless spots, which are located to

FIGURE 11. A female *Myrmekiaphila fluviatilis* in a defensive stance with chelicerae raised.

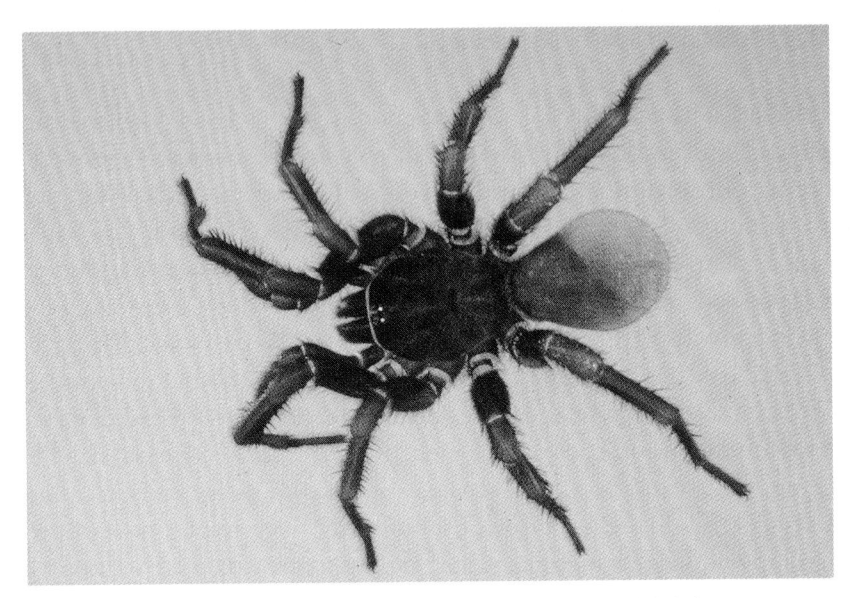

FIGURE 12. A large male *Myrmekiaphila fluviatilis.* Note the reddish legs.

each side of the midline. Lastly, the mature male will possess knobs on the medial surface of the tibia of the palps.

Ecology and Behavior: "Myrmekiaphila" means "ant lover." Many authors have declared that the spiders of this genus are frequently found within ant nests. We have collected numerous specimens throughout Tennessee, Alabama and Georgia and only one specimen has been found to be associated with an ant nest. The burrows are most consistently found on floors of temperate deciduous forests in which several inches of leaf litter cover their openings. Burrows are also likely to be found along ravines, hillsides, or bottom lands along streams. The burrows may be as shallow as four to five inches, especially for young spiders or burrows constructed in hard, rocky soils. However, in softer soils burrows have been as deep as eighteen inches, highly branched and they may house several spiders.

The *Myrmekiaphila* burrow can be located by first raking small sections of leaf litter. Because the trapdoor is most often a feeble piece of webbing woven with leaf matter, it will be cast away by the raking. While digging out and exposing the main shaft of the burrow, side branches leading from the main shaft were found. These side branches were also equipped with silken doors. The *Myrmekiaphila* burrow opening resembles that of the Ravine Trapdoor Spider, *Cyclocosmia truncata*. However, the burrow of the former is constructed diagonally to the surface of the ground while the *Cyclocosmia* burrow is perpendicular to the ground.

Myrmekiaphila spiders have never been known to be venomous to humans. However, due to their large size, aggressive behavior and fearless display of large chelicerae when disturbed, lead one to believe that their bite could be very painful, though not likely venomous.

Size: Length of female 24 mm; of male 19 mm.

FAMILY CTENIZIDAE—Trap-Door Spiders

These are large mygalomorph spiders, up to 25 mm in body length, and are distinguished by lacking abdominal tergites, claw tufts, and scopulae on female tarsi. They possess lateral rows of short spines on the tarsi and metatarsi of legs I and II. The thoracic furrow is strongly procurved. They also have biserially toothed chelicerae.They live in underground burrows lined with silk and closed with a trapdoor.

These mygalomorphs are found throughout the southern U.S. north to Illinois.. There are four genera with 14 described species. There are over 40 undescribed species in the genus *Ummidia* (Roth, 1993). Ctenizid spiders common in the southeastern U.S. include *Cyclocosmia truncata* and *Ummidia audouini*.

Family Ctenizidae—*Cyclocosmia truncata* (Hentz)—Ravine Trap-Door Spider

Identifying Characteristics: This primitive mygalomorph spider is perhaps one of the most distinctive and unusual spiders in the southeastern U.S. Its abdomen is bluntly truncated at the rear end and is covered by 48 to 52 longitudinal ridges on each side. It has two pairs of lungs. Color differences exist between the sexes. Females have a brown carapace and legs, with the anterior half of the abdomen being brown while the posterior half is black. Males have a black carapace, black legs and a light brown abdomen.

Ecology and Behavior: This unique spider was originally discovered in Alabama during the 1800's and was described as a species new to science by the famous arachnologist, Dr. Nicholas Marcellus Hentz, M. D. This trapdoor spider makes a vertical burrow capped by a thin silken, hinged

FIGURE 13. Female *Cyclocosmia truncata.* Note that the elongated palps resemble a fifth pair of legs, a condition seen in most mygalomorph spiders.

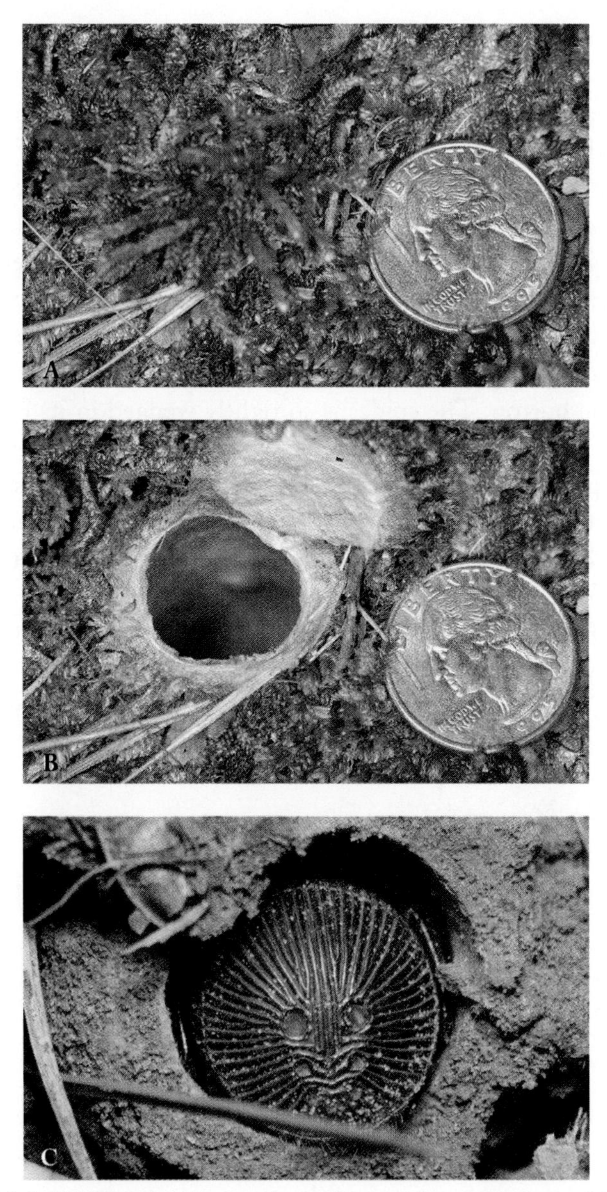

FIGURE 14 A-C. (**A**) The closed, moss-covered camouflaged trapdoor of *Cyclocosmia truncata* just to the left of the coin; (**B**) the same trapdoor of *C. truncata* (as seen in A) but now opened and seen to the left of the coin; (**C**) the circular, flattened rear end of the abdomen of *C. truncata* nearly plugging the entrance to its burrow.

trapdoor. The burrow is almost always found along steeply sloped banks of ravines in areas of hardwood forest floor covered by moist leaf litter. The deepest burrow that we have found was about 20 cm (about 8 inches). The entrance to the burrow depends on the size of the spider inhabiting it, but those built by large adult spiders usually are about 20 mm in diameter (about the diameter of a quarter). According to some authorities, the burrow diameter narrows slightly from the entrance to the bottom. When the spider is threatened by a predator, it crawls head downward in its burrow until its circular abdomen completely fills the circular passageway and plugs it tightly. It literally creates a "false bottom" to the burrow. The spider is so tightly wedged in its burrow that some experts claim that it cannot be removed from above without tearing the spider's body apart. We have unearthed a large female that guarded her egg sac beneath the "false bottom" that she created in her burrow. The egg sac was wedged two inches above the bottom of the burrow, possibly to provide drainage.

C. truncata is a long-lived species which may live up to twelve years. Males may be found wandering about in search of females during late Fall. This spider is found only in north Georgia, north and central Alabama, and Tennessee (Gertsch and Platnick, 1975).

Size: Length of female 33 mm; of males 19 mm.

Family Ctenizidae—*Ummidia audouini* (Lucas)

Identifying Characteristics: *Ummidia audouini* is one of the largest of the trapdoor spiders of the southeastern United States. The specimen appearing in the photograph was collected from east central Alabama and has a black cephalothorax, chelicera, sternum and legs. Specimens described from outside the southeast, i.e., Kansas, had a lighter color which has been described as "chestnut brown" or "dark amber." The Alabama specimen has the typical dark gray abdomen with a slightly wrinkled integument. The legs are short and have concentric white bands at the ends of the femur, tibia, and tarsi. The presence of a deep shiny depression of the dorsal surface of tibia III is a positive identifying feature of this spider.

Ecology and behavior: The burrow of *Ummidia audouini* is typically found along the sides of ravines in deciduous forests. Where we have found

FIGURE 15. A large female *Ummidia audouini* showing her black shiny carapace and huge abdomen.

Ummidia burrows, we have also found *Cyclocosmia* and *Myrmekiaphila* burrows just a few feet away. The *Ummidia* burrow is distinctive in being shallow, rarely exceeding a depth of five inches. The spider conceals its burrow opening with a sturdy, silken, hinged trapdoor. The *Ummidia* burrow is covered by a superbly camouflaged trapdoor that can be easily overlooked even by the trained eye. The *Ummidia* burrow opening has a diameter of approximately one inch, which can be two to four times the size of burrow opening of other trapdoor spider genera. The *Ummidia* burrow is completely lined with a tough silken sac. The integrity of these sacs is such that if grabbed at the edge of the mouth of the burrow, it sometimes can be pulled out of the ground, intact! We have witnessed the fast work of a female *Ummidia audouini* dig her burrow and construct the burrow lining and trapdoor in a single evening.

While *Ummidia* is a burrowing spider, it can be forced out of its burrow by special circumstances. Wandering males can be found in July and August in search of a mate. During hot, dry summers female *Ummidia* spiders have been found drowned in watering bowls for pets, and even in outdoor swimming pools.

Size: Length of female 28 mm; of male 15 mm (Kaston, 1978).

B. SUBORDER ARANEOMORPHAE

Some authors prefer to use the suborder Labidognatha instead of Araneomorphae. Either classification is acceptable. These spiders differ from those tarantula-like spiders of the suborder Mygalomorphae by having chelicerae that project downward, or obliquely downward and forward, with the fangs hinged so that they move in a transverse plane. They usually have one pair of book lungs, and either a single median tracheal spiracle or a pair of spiracles situated at a varying distance between the lung slits and the anterior spinnerets depending upon the species.

FAMILY HYPOCHILIDAE—Lampshade Weavers

These interesting spiders are distinguished from all other araneomorph spiders by having four book lungs. The labium and sternum are fused. The cheliceral fangs are obliquely angled. The legs are long and slender similar to those of cellar spiders (Roth, 1993). These spiders weave a web that is shaped somewhat like that of a lampshade. The web is attached to the underside of overhanging ledges, the roof of rock bluff shelters or in ceilings of shallow caves. The lampshade webs look like tiny lampshades whose small ends have been glued to the ceilings of rock ledges and caves. The spider hangs upside down on the rock ceilings surrounded by the lampshade web. In the eastern U.S., lampshade weavers are found in the Appalachian Mountains from West Virginia south into north Alabama. Western lampshade weavers have been found in Colorado, New Mexico and California. *Hypochilus* is the only genus in this family and it contains 8 species (Roth, 1993). Taxonomic reviews of this family were written by Forster et al. (1987) and Gertsch (1958; 1964).

FIGURE 16. The "lampshade" web of *Hypochilus* attached to the underside of a rock ledge in Little River Canyon of North Alabama.

Family Hypochilidae—*Hypochilus thorellii* (Marx)—Lampshade Weaver

Identifying Characteristics: *Hypochilus* is a unique spider which somewhat resembles the cellar spiders of the family Pholcidae by having an elongate body and long and slender legs. The body is grayish yellow with irregular dark purplish brown blotches (Kaston, 1978). Members of this genus are similar to the tarantulas in that they have their venom glands located entirely within their chelicerae (Kaston, 1978). *Hypochilus* is best distinguished in the field by its unique web and highly specific habitat. Its web consists of irregular silk mesh constructed in the shape of a lampshade. Its habitat is almost always found on the undersurfaces of overhanging rock ledges or in cave entrances usually near a stream or water falls. It is best found by searching for its distinctive lampshade-like webs which hang down on the undersides of rock ledges. The spider sits upside down in the top of the lampshade web while resting against the undersurface of the ledge.

FIGURE 17. *Hypochilus thorellii* rests against the undersurface of a rock ledge. It is surrounded by its lampshade-like web which is attached to the rock ledge and hangs down around the spider.

Size: Length of female 14 to 15.5 mm; of male, 10 to 11 mm (Kaston, 1978).

FAMILY FILISTATIDAE—Crevice Weavers

Filistatids have a cribellum and a short calamistrum. The labium is fused to the sternum. The basal portions of the chelicerae are fused along their inner margins. The eight eyes are closely grouped on a rounded prominence.

These spiders are well-known for building a tubular retreat in cracks and crevices of rocks and in the boards of old houses. The familiar radiating lines of hackled silk lead into the silk lined tube.

The family occurs commonly throughout the southern U.S. and is represented by three genera and seven described species. There are perhaps seven undescribed species (Roth, 1993). One species, *Kukulcania hibernalis* (Hentz) is one of the most common spiders in the eastern U.S.

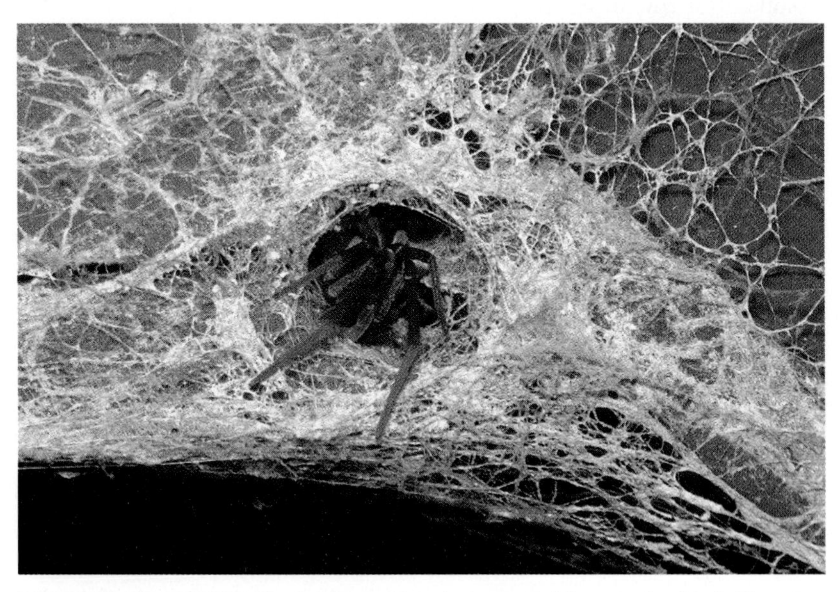

FIGURE 18. Female *Kukulcania hibernalis* in her retreat. Note the hackled silk webbing characteristic of cribellate spiders.

Family Filistatidae—*Kukulcania hibernalis* (Hentz)—Crevice Spider, Southern House Spider

Identifying Characteristics: This spider has been previously known as *Filistata hibernalis* Hentz. This is one of the most common spiders in the eastern states that has a calamistrum and a cribellum. This is a rather large spider and is frequently found around most human dwellings in the south, especially abandoned homes and barns. The spider varies a great deal in coloration, being dark gray to black in some geographical areas, and light beige to brown in other regions. The most common coloration in southeastern populations is homogeneous light beige or gray on the abdomen and a correspondingly slightly darker color on the carapace. The eyes are clustered in a group which forms a distinctively darker area on the head. In the lighter colored male, the markings on the carapace are reminiscent of the "violin" marking on the poisonous brown recluse spider, *Loxosceles reclusa*. Because of this marking and its homogeneously beige color over the abdomen and carapace, the male has often been mistaken for the brown recluse. The pedipalps of the male are unusually long and simple in structure. The adult male lacks the calamistrum. In the female, the calamistrum is near the base of the fourth metatarsus and is relatively short. The legs are very long, with the first pair being nearly twice as long as the body.

Ecology and Behavior: The webs of these spiders are very conspicuous, usually seen on the sides of unkept buildings, in basements and in attics. The web is usually circular and surrounds the opening of the retreat of the spider. The webs are most often constructed around the crevices and cracks usually associated with old window sills, doorways, walls and baseboards. It is in these holes that the spider makes its hiding place and its circular web fans outwardly from the opening of the retreat. The spider is relatively sedentary and is rarely seen. If an insect is placed on its web , the spider will usually immediately rush out of its hiding place, attack the prey and carry it into its retreat. The web has a unique hackled appearance. Comstock (1940) found that the spider produces four different kinds of silk in producing the web. According to Kaston (1978),

FIGURE 19. Female of *Kukulcania hibernalis.*

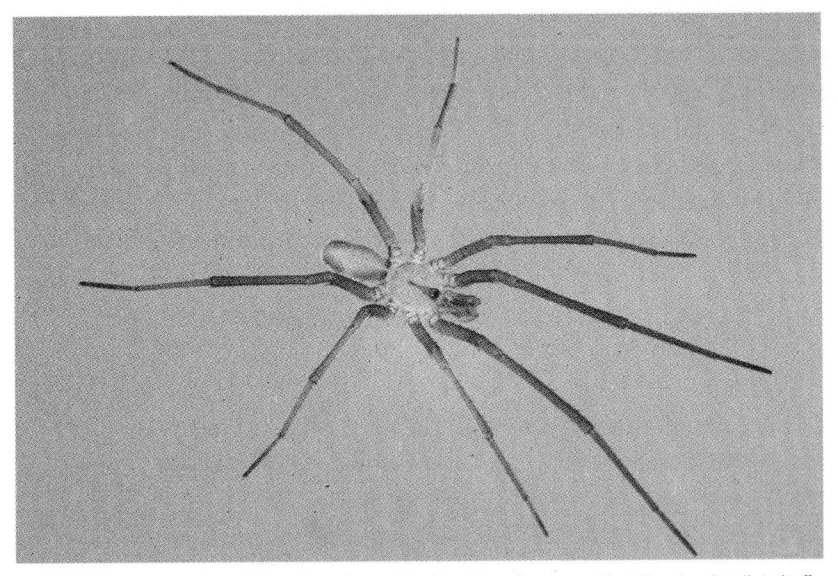

FIGURE 20. Male of *Kukulcania hibernalis* showing the superficially similar "violin" pattern on the carapace that often leads some to mistake this spider for the brown recluse.

females may live up to eight years. This species is distributed throughout the southern U.S. states.

Size: Length of female 13 to 19 mm; of male 9 to 10 mm (Kaston, 1978).

FAMILY SICARIIDAE—Sixeyed Sicariid Spiders

Some members of this family were long placed in the family Scytodidae, but were later transferred to the family Loxoscelidae (Gertsch and Ennik, 1983). More recently these spiders were placed into the family Sicariidae primarily on the basis of spinneret morphology (Platnick et al. 1991). These spiders are characterized by having the following features: 6 eyes in a strongly recurved row of 3 dyads; carapace relatively flat above and unmodified; tarsi with 2 claws, sternum pointed behind; spiracular furrow one-sixth the distance from spinnerets to epigastric furrow; outer side of chelicerae with stidulating files; and, femur of palpus with one short black stridulating pin near base.

These spiders make small, irregular webs under logs and stones, but may also take up residence indoors where they hide in basements and closets under clothing, inside shoes, beneath paper and other items. These spiders have become well-known for their ability to envenomate humans and cause extensive tissue necrosis in the area of the bite.

The family has one genus, *Loxosceles,* with 13 species (Roth, 1993). It is distributed throughout the southern states from Georgia to California. The brown recluse spider is a typical representative of this family.

Family Sicariidae—*Loxosceles reclusa* (Gertsch and Mulaik)—Brown Recluse or Violin Spider

Identifying Characteristics: The brown recluse spider is one of the two dangerous spiders to occur within the eastern U.S., the other species being the black widow spider. Fortunately, as with the black widow

FIGURE 21. Female Brown Recluse with a dark brown violin-shaped pattern on the cephalothorax.

spider, the brown recluse is readily identifiable. It has a light beige cephalothorax which has a dark brown violin-shaped mark. Some people call this spider the "violin spider". The abdomen is homogeneously beige to light brown in color. This spider has relatively long, thin legs with the greatest distance between opposing tarsi (toes) being about 50 mm (2 inches). Leg I, in females, is almost 4.5 times as long as the carapace; in males, this leg is 5.5 times the length of the carapace.

Ecology and Behavior: The brown recluse is relatively common in the southeastern U.S. Their habitat is often indoors in dark and dry basements, closets, toes of shoes, old rolled up newspapers, in trunks, and in folded and hanging clothes which have not been used for some time. When outdoors, they live under logs, under loose tree bark and under rocks. In these areas, they construct small irregular webs. The spider is found from Ohio south to Georgia and west to Texas and Nebraska (Kaston, 1978).

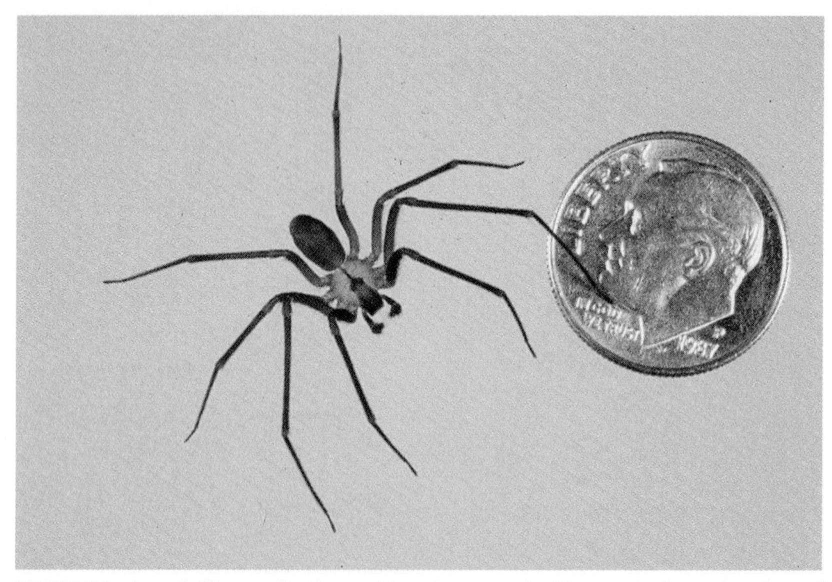

FIGURE 22. A male Brown Recluse spider photographed by a coin for estimation of size.

Brown Recluse Bites: This spider is venomous and its capture should not be attempted except by professionally trained biologists. Most people are accidentally bitten when they put on old shoes or old clothing items that have not been recently worn. The brown recluse venom is hemolytic, differing from that of the neurotoxic black widow spider. The toxin of the brown recluse can cause tissue death at the bite site and destruction of blood cells. The bite can be mild to serious, and occasionally fatal. Within an hour after the bite, a painful edematous swelling occurs, and skin surrounding the bite swells into a fluid-filled blister. This is followed with nausea, abdominal cramps, pain and fever. As the edema subsides at the site of the bite, bleeding, ulceration and gangrene may occur, with necrosis (tissue death) causing the outer skin layers to slough away. A few deaths have been reported, usually in small children, due to hemolytic anemia and renal failure. Although there is no antidote known, some physicians administer corticosteroid therapy to reduce the inflammatory response and analgesics for pain. Some people may need skin grafts due to severe tissue death. To help avoid being bitten, one should shake out shoes and clothing before use; eliminate collections of old newspapers and unused boxes; clean beneath and behind furniture; remove spiders, spider webs and any spider egg cases from all living and storage areas. Local pest control companies may be called to remove these dangerous spiders from around your house and yard.

Size: Length of female, 9 mm; of male 8 mm (Kaston, 1978).

FAMILY SCYTODIDAE—Spitting Spiders

The "spitting spiders" are distinguished by having 3-claws on the tarsi and 6 eyes on the head; eyes arranged into 3 dyads in a strongly recurved row; the carapace strongly arched posteriorly; 2 sclerotized ridges behind epigastric furrow in female; colulus present; chelicerae fused basally;

FIGURE 23. Female *Scytodes sp.* The spitting spiders carry their eggs with their chelicerae.

sternum truncated behind; spiracular furrow one-fourth the distance from the spinnerets to epigastric furrow.

These spiders have enormous poison glands and capture their prey by spitting a gummy secretion over them, fastening them to the substrate. They are slow-moving, nocturnal spiders which are often found in dark corners, basements and closets. They carry their eggsacs by their chelicerae.

The family has one genus, *Scytodes,* with 6 species and is distributed throughout most of the U.S. (Roth, 1993). *Scytodes thoracica* is a relatively wide-spread species and will represent this family.

Family Scytodidae—*Scytodes thoracica* (Latreille)—Spitting Spider

Identifying Characteristics: This strange spider is believed to have been imported from Europe (Emerton, 1902:131). The cephalothorax and

FIGURE 24. *Scytodes thoracica* is the most common spitting spider in the eastern U.S.

abdomen are round and are almost the same size in diameter. The ground color of the body is yellow with a definite pattern of black spots. The yellow legs have three dark rings on the femora and tibiae. The front of the head is extended into a truncated projection which extends forward over the chelicerae. There are only 6 eyes, arranged into three pairs. The chelicerae are small and fused at the base. A colulus is present and relatively large.

Ecology and Behavior: This spider is unique in that it catches its prey by spitting a mucilaginous secretion over it which sticks the unlucky victim to the substratum. The venom gland is very large and has two distinct regions. The smaller region near the anterior end of the gland secretes venom as in other spiders. The much larger posterior region secretes the mucilaginous substance for spitting at prey. These are very slow-moving spiders found in ground litter, under rocks, and more often in cellars and closets. These spiders are usually not associated with webs. Mature specimens have been taken during June through October. Mating usually occurs during September and October (Kaston, 1948). This species occurs throughout the eastern U.S. into Canada (Kaston, 1978).

Size: Length of female 4 to 5.5 mm; of male 3.5 to 4 mm (Kaston, 1978).

FAMILY PHOLCIDAE—Cellar Spiders

These spiders have extremely long and thin legs and are often confused with another group of arachnids called daddylonglegs (Opiliones or harvestmen). Harvestmen can be distinguished from these cellar spiders by having only one body region, whereas the spider has two. Most cellar spiders have 8 eyes, but some have only 6. The anterior median eyes are the smallest (if present) while the other eyes are clustered in two triads. These spiders are common in dark places such as basements where they make highly irregular webs which they vibrate when disturbed. They carry their egg sacs with the chelicerae. About 36 species occur in North America (Roth, 1993).

Family Pholcidae—*Pholcus phalangioides* (Fuesslin)—Longbodied Cellar Spider

Identifying Characteristics: This spider has extremely long legs like those of a daddylonglegs. According to Emerton (1961), "The body is quarter of an inch long, and the longest legs two inches." The color of the carapace is usually pale gray except for a dark coloration around the eyes and a distinctive dark gray marking in the middle of the carapace. The abdomen is also pale gray except for a translucent middorsal line. The elongate abdomen is about three times as long as wide. Unless the abdomen is distended with eggs, its sides are relatively straight. The cephalothorax is nearly round and flat behind. The part of the head surrounding the eyes is raised and in the males is separated at the sides from the rest of the head (Emerton, 1961). The anterior eyes are dark and are the smallest of the eight eyes. The other six eyes are light in color and are arranged in two elevated triads.

FIGURE 25. A cellar spider, *Pholcus phalangioides*, hanging in its typical upside down position in its web. Note the young spiderlings nearby that have not yet left their mother's web.

Ecology and Behavior: This spider is common in the U.S. and Europe, and perhaps represents an imported species. It inhabits dark basements and cellars. It has also been found in caves and beneath rock overhangs where there is little light. It constructs a large web consisting of an irregular maze of threads. The spider hangs upside down in the web, and when threatened, it swings back and forth so rapidly that it can hardly be seen. The female encases her eggs in an egg sac that is so thin that the eggs may be seen through the membraneous covering. The egg sac is carried about in the spider's chelicerae until the young hatch. The cellar spider may live up to three years, two as an adult . According to Kaston (1978) this is the commonest species of cellar spider in the U.S.

Size: Length of female 7 to 8 mm; of male 6 mm (Kaston, 1978).

Family Pholcidae—*Physocyclus globosus* (Taczanowski)

Identifying Characteristics: Cellar spiders of the genus *Physocyclus* are distinguished by a combination of three primary features. First, as the

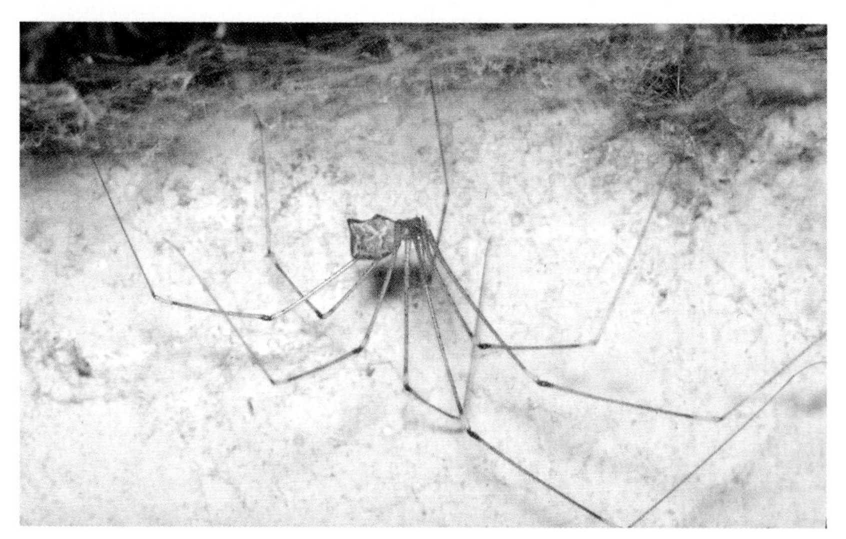

FIGURE 26. Female *Physocyclus globosus* suspended in her web.

specific name implies, the pale gray and speckled abdomen is rounded, being as long as wide and as high. Second, among the long spindly legs, the femur of Leg I is at least twice as long as the cephalothorax. Furthermore, the femur of Leg IV is at least as long as the femur of Leg I. Each leg is light yellow with dark rings on the distal ends of the femur and tibia. Third, the posterior row of eyes are slightly recurved, such that the lateral eyes are further posterior than the median eyes.

Ecology and Behavior: *P. globosus* spins a loose irregular web, similar to *Pholcus phalangioides*. When disturbed the spider shakes so rapidly that it appears to spin. Generally, where the former species is found, so can the latter be found. They inhabit cellars and sheds of the warmer regions of the country. Their range extends from Georgia and Florida west to California (Comstock, 1912).

Size: Length of female 4.7 mm; of male 3.7 mm (Kaston, 1978).

FAMILY SEGESTRIIDAE—Segestriid Spiders

These spiders are distinguished by having only 6 eyes and a conspicuous pair of tracheal spiracles behind the epigastric furrow. The cheliceral margins are toothed and the third pair of legs are directed forward. The tarsi bear 3 claws.

They build a silken tubular retreat in the cracks of rocks and trees and under stones and bark. They emerge at night and feed around the entrance of their burrow.

There are about two genera and 6 species with the family represented throughout most of the U.S. (Roth, 1993). Previous to the work of Forster and Platnick (1985), arachnologists had considered Segestriidae and Dysderidae to be subgroups within the family Dysderidae. Forster and Platnick (1985) split the Segestriidae off into a family distinctive from Dysderidae. The most common representative of this family in the eastern U.S. is *Ariadna bicolor*.

FIGURE 27. Entrance into the silken retreat of *Ariadna bicolor.*

Family Segestriidae—*Ariadna bicolor* (Hentz)

Identifying Characteristics: The cephalothorax is relatively narrow and elongated. It is yellowish-brown. The abdomen is elongated and narrow, sometimes being slightly oval. It is purplish brown on the dorsum and venter, paler along the sides. Its legs are yellowish-brown. It has six eyes in three pairs grouped closely together. The side pairs are separated by their diameter from the middle pair (Emerton, 1961). The first, second and third pairs of legs are directed forward. The first pair of legs is the longest and stoutest. The tarsi bear three claws. The sternum is twice as long as broad and lacks the lateral extensions which are characteristic of the closely related *Dysdera crocata.*

Ecology and Behavior: This spider is distinguished by its remarkable habits of constructing a unique silken tubular retreat in the cracks of trees, rocks, and under the bark of trees and stones. In Gaffney, SC, we studied the tubular retreat of this spider. It was constructed in dense

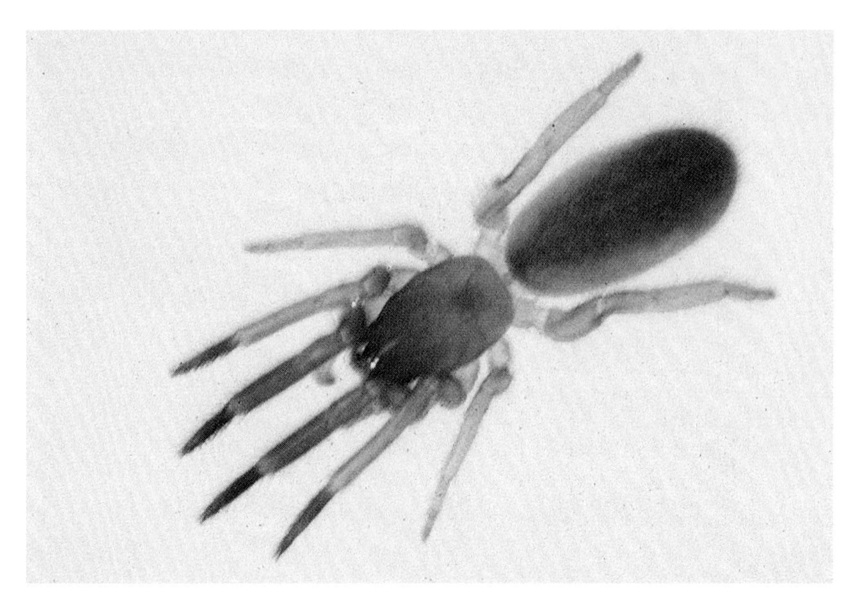

FIGURE 28. *Ariadna bicolor* in its typical posture of legs I, II and III directed forward.

deciduous woodlands along the side of a fallen, partially decayed trunk of a tree which was about eight inches in diameter. The tree trunk was two-thirds covered by leaf-litter. The entrance to the tube was conspicuous when seen from above. It consisted of a hole about 10 mm in diameter and this was surrounded by silken threads forming a sort of collar-like rim around it. The rim and hole extended to about 30 mm in diameter. Radiating from the rim were fine threads arranged in such a manner that any insect walking over them would communicate its presence to the predaceous spider inside the tube . When the leaf-litter was removed from around the tree-trunk, a white silken tube was seen attached to the side of the tree trunk and it extended a length of about 45 mm. When the tube was probed, the spider ran out immediately with amazing speed and disappeared into the leaf litter of the forest floor. We have also collected *Ariadna* beneath the bark of living pine trees. According to Comstock (1940), "The spider waits within the tube with six of its eight legs projecting forward ready to make a leap. The touching of one of the trap lines by an insect results like the touching of the spring of a jack-in-the-box. The spider comes forth with amazing swiftness, seizes the unlucky insect, and retreats with it instantly to its lair." About 15 relatively large eggs are laid within the tube in a spherical mass and are not enclosed in an egg sac (Comstock, 1940). Beatty (1970) studied the taxonomy of the genus *Ariadna*. The spider is distributed from New England south to Florida and westward along central and southern states to Colorado and California (Kaston, 1978).

Size: Length of female 6.1 to 15 mm; of male 5.4 to 10.6 mm (Kaston, 1978).

FAMILY DYSDERIDAE—Dysderid Spiders

This small family of ecribellate spiders is distinguished by having only six eyes. The tarsi bear a pair of claws which are underskirted by a tuft of bristles. Ventrally the sternum possesses lateral extensions between each coxae. The labium is deeply notched and fused to the sternum. The chelicerae are very large and project forward, and the fangs are almost parallel. The family has one species, *Dysdera crocata* in the eastern U.S.

Family Dysderidae—*Dysdera crocata* (C.L. Koch)

Identifying Characteristics: *Dysdera crocata* is the only species of the family found in the United States and is widely distributed east of the Mississippi River. It is easily recognized by its orange cephalothorax, chelicera and legs. The chelicerae are greatly enlarged and are held extended from the cephalothorax. On the chelicerae are combs of long bristles and fangs that are nearly as long as the chelicerae. The chelicerae and fangs in the photographed specimen appear to be so large that they seem like they would be as much of a liability as an asset. Ventrally, the labium is joined by an enlarged anterior plate from the first coxae. The abdomen is gray to dirty white without distinguishing markings.

FIGURE 29. A female *Dysdera crocata,* showing her enlarged chelicerae and fangs

Upon first appearance this spider resembles a mygalomorph due to the elongate hairless abdomen, enlarged chelicera, and two pairs of lung openings. The chelicera operate on the horizontal plane as do the labidognath (araneomorph) spiders.

Ecology and Behavior: These spiders live under stones and loose tree bark. The photographed specimen was collected from the underside of a sheet of abandoned roofing tin, where the spider had sought a dark and humid nesting area. This spider does not spin a web but constructs a tough oval silken retreat hardly larger than the spider itself. These spiders are nocturnal, leaving their silken sanctuary as they hunt for their prey. Supposedly they specialize in feeding on pill bugs (isopods).

Size: Length of female 11 to 15 mm; of male 9 to 10 mm (Kaston, 1978).

FAMILY MIMETIDAE—Pirate Spiders

The family name Mimetidae was derived from the Greek *mimetos* which means to be imitated or copied. Originally these spiders were mistakenly thought to build a double web, similar to the irregular webs of the spiders of Theridiidae or the orb webs of the Araneidae. While they can be found in both types of webs, the mimetid spiders do not construct them. Rather, they are parasites that infiltrate the host's web and feed on resident spiders.

The first two pairs of legs of the mimetid spiders are twice as long as the third and fourth legs. Distinctive from all other ecribellate spiders, the tibia and metatarsi of leg I and II are armed with a series of long spines, interspersed with shorter spines. The chelicera are large and armed on the retromargin of the fang furrow with long spines. The eyes are positioned on the front of the head and the clypeus is much reduced.

Mimetidae is a small family of two genera and fourteen North American species (Roth, 1993). Chamberlin (1923) gave a review of the North

FIGURE 30. A pirate spider, *Mimetus puritanus,* loiters about a mud dauber nest stealing the dauber's captured spider prey.

American species of the genus *Mimetus*. We represent the family with the most common pirate spider of the eastern U.S., *Mimetus puritanus*.

Family Mimetidae—*Mimetus puritanus* (Chamblerlin)

Identifying Characteristics: The ground color of this pirate spider is pale yellow. While many of the markings are variable, a double V-shaped mark on the white cephalothorax is distinctive. S-shaped black markings extend the length of both sides of the abdominal dorsum. Three or four lines traverse the posterior end of the abdomen. As in the photographed specimen, many individuals have white spots on the abdomen, anteriorly. Scattered over the entire dorsal surface of the abdomen and legs are numerous bright red pin-point dots. The legs are light yellow with dark rings at the ends of the joints.

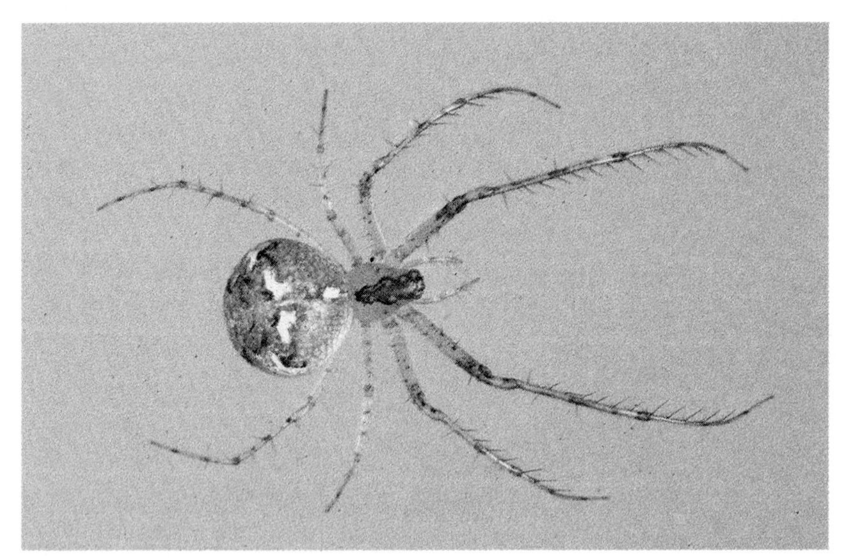

FIGURE 31. This extraordinarily pale specimen of *Mimetus puritanus* reveals the typical body markings and spines on the legs.

Ecology and Behavior: Like other mimetids, *M. puritanus,* usually does not spin a snare of its own. Rather, it invades the webs of other spiders where it bites and paralyzes the inhabitant. It sucks the host dry of hemolymph through the legs, one leg after the other. This pirate spider lives in warm and dry areas on bushes, fences, and around the eaves of houses. It can be found inside houses where it invades the webs of the common house spider, *Achaearanea tepidariorum.* Females deposit an oblong egg sac with tapered ends in the web of its victim. According to Kaston (1978), *M. puritanus* is distributed from New England and adjacent Canada south to Georgia and west to Wisconsin and Kansas.

Size: Length of female 5 to 5.6 mm; of male 4 to 4.5 mm (Kaston, 1978).

FAMILY OECOBIIDAE—Flatmesh Weavers

These are very small spiders ranging from 0.9 to 3.2 mm in body length. They have a cribellum which may be rudimentary. The males lack a calamistrum. They may be distinguished from all other spiders by their large, hairy anal tubercle. The carapace and sternum are wider than long and rounded in appearance. They have three claws. They have a long, upturned and pointed apical segment of the posterior spinnerets.

They construct small, somewhat flattened, silken retreats usually less than the diameter of a penny, over crevices and depressions in walls, in the mortar joints of bricks, or on the undersides of rocks (Roth, 1993; Shear, 1970).

Oecobiids are widespread throughout the U.S., with several species introduced. There are two genera and 7 species.

Family Oecobiidae—*Oecobius parietalis* (Hentz)

Identifying Characteristics: The ground color of both the cepahlothorax, abdomen and legs is pale yellow. This creates a high contrast with the brown to brownish-black pigment which forms dots, lines and blotches

FIGURE 32. A camouflaged *Oecobius parietalis* in a mortar seam of a brick wall. The spider is in the middle of the photograph at the tip of the arrow.

on the cephalothorax and abdomen, and rings on the legs. The cephalothorax has a distinctive brownish-black "Mickey Mouse head and ears" pattern , with the "head" in the center of the carapace and the "ears" at the anterior end. Along each side of the carapace is a row of three brownish-black dots. There is much variation among individuals within a single population relative to the color pattern on the abdomen. However, most females have a marginal black band on the anterior one-half of the abdomen. This band follows the anterior contour of the abdomen. Otherwise, the anterior one-half of the abdomen is largely devoid of dark pigment. The posterior one-half of the abdomen is heavily marked with dark blotches, dots and lines. Six to seven light brown annuli encircle each leg. The carapace and sternum on this and the closely-related *Oecobius annulipes* are wider than long. Females have a cribellum and a calamistrum but these may be rudimentary in the males.

Ecology and Behavior: This is a very small spider, measuring only 3.0 mm in length. It is an inconspicuous species that is rarely noticed by most

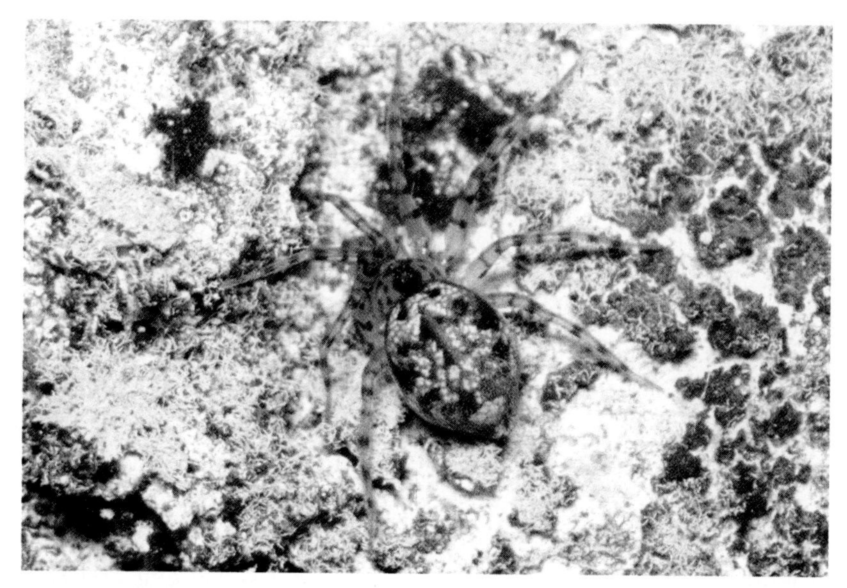

FIGURE 33. Closeup of *Oecobius parietalis* on a mortar seam of an old brick wall.

people. It makes its home in the crevices on the outside of buildings and walls, and sometimes, within buildings. These spiders are attracted to human-made structures and are truly domestic spiders. According to Comstock (1912), this fact is suggested by the generic name, which is derived from the Greek "oikobios" which means "living at home". We have found this spider most often on protected outside walls and stairwells of brick buildings where the spider lives beneath small flat silk webs constructed in the recessed mortar joints between bricks. Here, the spider makes a sheet-like web less than one-inch in diameter. The spider usually lies in wait just beneath this sheet and the outline of the animal can often be seen through the thin webbing. When disturbed, it will leave the cocoon-like retreat and run with great speed, being very difficult to capture.

Size: Length of female 3.0 mm; of male 2.6 mm.

FAMILY ULOBORIDAE—Hackled Orbweavers

This is the only North American spider family which lacks venom glands. The females have a well-developed pair of humps near the anterior end of the abdomen. While females usually have a well-developed cribellum and calamistrum, some males lack these structures (Opell, 1979). There is often a feathery group of hairs on the distal end of the tibia I. According to Roth (1993), the family can further be distinguished from other cribellates by having a prominent, ventral row of short, heavy setae from mid-metatarsus to the tip of tarsus IV.

The tiny webs of uloborids are horizontally oriented, which distinguish them from most other orbweavers. The common name for this family comes from the messy or hackled appearance of the individual web strands, caused by the combing of the web with the calamistrum.

There are seven genera and about 15 species (Roth, 1993). The family is distributed throughout most of the U.S. (Muma and Gertsch, 1964; Opell, 1979).

FIGURE 34. A small horizontally placed web of a feather-legged spider. A heavy band of zig-zag silk forms a stabilimentum which crosses the center of the web.

Family Uloboridae—*Hyptiotes cavatus* (Hentz)—Triangle Spider

Identifying Characteristics: The triangle spider may vary in general overall background coloration from pale yellow to gray, to dark brown or black. There are no distinctive color patterns against the background coloration. The carapace is usually brown with an indistinct light median stripe which becomes more delineated in the median eye quadrangle (Muma and Gertsch, 1964). Its body is covered in short, semi-erect, black, brown, and white hairs and scales. The abdomen, which is variable in its base color, is marked dorsally with dark spots over small tubercles. Dark lines often connect each pair of tubercles. There is an interrupted, dark, median dorsal stripe. The sides have indistinct dark bars and darker shading around the spinnerets. Most of the venter is brown to black in color. The legs and palpi are brown. The abdomen of the female has a double row of rounded tubercles on each side of the dorsum. The tubercles are not as prominent in the smaller male.

FIGURE 35. The tiny triangle spider, *Hyptiotes cavatus*, camouflaged on a twig.

Ecology and Behavior: The triangle spider, like other uloborids, are the only known cribellate spiders that spin capturing webs in the form of orbs or portions of orbs (Muma and Gertsch, 1964). *Hyptiotes* spins a sector of about 45° of an orbweb and places it in a vertical position in the form of a triangular snare. *Hyptiotes* constructs the web so that the line upon which the four radii converge is attached to a twig. The spider usually positions itself, dorsum down, close to the twig so that it resembles a twig bud. The spider holds on to the twig and, at the same time holds onto the snare line and stretches it tight. When a flying insect gets caught in the triangular portion of the web, the spider releases the tension on the snare line and this sends waves of loose snare threads over the prey and entangles it even more. *Hyptiotes* and other uloborids are also known as the hackled-band orbweavers.

Size: Length of female 2.3 to 4 mm; of male 2 to 2.6 mm (Kaston, 1978)

Family Uloboridae—*Uloborus glomosus* (Walckenaer)—Feather-legged Orbweaver

Identifying Characteristics: The females have leg I about 4.5 times as long as the carapace. This leg bears a distinctive brush of hairs at the distal end of each tibia. Males lack the brush of hairs on leg I. The posterior row of eyes is recurved to such an extent that a line drawn along the front edge of the laterals does not touch the medians. On the dorsum of the abdomen are a pair of protuberances, or humps, located about one-third the distance from the anterior end. The general body coloration is grayish brown.

Ecology and Behavior: This spider, like other members of its genus, always spins a complete horizontal orb web, 100 to 150 mm in diameter (Kaston, 1978). The web is provided with a stabilimentum. The web is usually placed in bushes or tree branches a few feet from the ground. According to Moulder (1992), the spider may overwinter as a juvenile and complete its maturation the following spring. The female mates in

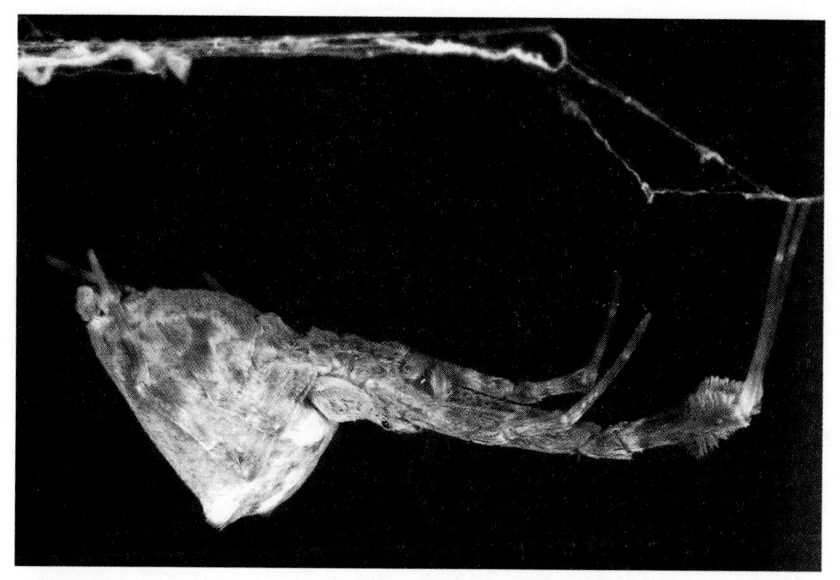

FIGURE 36. A female *Uloborus glomosus* displaying the "feathers" on the tibia of leg I.

early summer and produces egg sacs in June and July. The species is distributed from southern Canada to Florida and westward to Texas and Nebraska.

Size: Length of female 2.8 to 5.0 mm; of male 2.3 to 3.2 mm (Kaston, 1978).

FAMILY THERIDIIDAE—Comb-Footed Spiders

Theridiid spiders have been appropriately called the comb-footed spiders since Comstock proposed the name in 1912 (Comstock, 1912). Its comb-foot consists of a row of curved bristles on the tarsi of leg IV. The comb is used to cast semi-liquid silk from the spinnerets around prey that become entangled in its web. The web is a most disordered arrangement of framing strands that are intersected by short diagonal strands at most any angle. At the center of the irregular web, the strands increase in density and frequency of branching. Here the spider often will suspend itself in an inverted position where it remains rather sedentary as it patiently waits for prey to become entangled in its silken snare. Females will hang one or more egg sacs in the web periphery.

Many, but not all, theriidids have a globular shaped abdomen with spinnerets placed ventrally rather than posteriorly. The legs are moderately long and spiny, except on the tarsi and metatarsi of some species. The theridiid combs on tarsi IV should not be confused with the calamistrum located on metatarsi IV of the cribellate spiders. The chelicera of most theridiid species are small and lack marginal teeth.

Almost everyone is familiar with the theridiid spiders and webs from our experience with the common house spider, *Achaearanea tepidariorum*. This spider takes the liberty to keep its own house in the corners of our homes, basements, barns, and any dry structure. The vast majority of the theridiid species are more reclusive of humans, seeking refuge among plants and forest litter. The widow spiders seek out dry

FIGURE 37. Female Common House Spider, *Achaearanea tepidariorum*.

areas under logs and rocks. The *Argyrodes* are more inclined to take up commensal residence in the webs of other spiders. This family is very large and contains some 27 genera and 232 species (Roth, 1993). We represent this family with ten of the most commonly encountered genera from the eastern U.S.

Family Theridiidae—*Achaearanea rupicola* (Emerton)

Identifying Characteristics: In general body coloration, this little spider very closely resembles the young of the house spider, *Achaearanea tepidariorum*. The body is dirty gray with darker grayish to black spots and stripes. The legs have alternating light and dark rings. In the middle of the abdomen is a small hump which terminates in a tiny, but conspicuous bump or tubercle. The tubercular region is black toward the front and white toward the rear.

FIGURE 38. Female *Achaearanea rupicola* bearing the unique dark pimple on the posterior abdomen.

Ecology and Behavior: This is a small spider which constructs a small irregular-shaped web characteristic of most theridiids. The web is usually constructed around houses. Here, they may be found to inhabit water meter boxes, abandoned water pipes, old buckets and cans, the holes in cinder blocks, around rocks, and beneath boards. We found several of these spiders living adjacent to one another but separated within partitioned sections beneath a manhole cover. Within these sections, each spider had built an irregular web and had made a retreat camouflaged with leaves and debris. The female usually stays in this retreat and even lays her discoidal-shaped, white to brown egg sac attached to the inside. The female tenaciously guards her egg sac when disturbed. According to Emerton (1961) the webs of this species often contain grains of sand which look as if they were place there by the spider, because sand falling into such a web would obviously fall through without sticking to the threads. According to Kaston (1978) this spider is distributed from New England and adjacent Canada south to Alabama and west to the Mississippi River.

Size: Length of female 1.8 to 2.9 mm; of male 1.4 to 2 mm (Kaston, 1978).

Family Theridiidae—***Achaearanea tepidariorum*** (C. L. Koch)—Common House Spider

Identifying Characteristics: The ground color of the body varies from gray to a dull tan. The carapace is yellowish brown. The abdomen varies from dirty white to brown with a series of closely spaced indistinct gray chevron-like markings on the posterior end just above the spinnerets. Often, there is a triangular black mark near the middorsum of the abdomen just anterior to the chevrons. The legs are yellowish with gray annular rings at the ends of the segments. The abdomen of the female is large, inflated and high. The sternum is heart-shaped and varies from gray to dark brown. The legs of the male are orange and relatively longer than those of the female. The male's abdomen is less swollen and not as elevated.

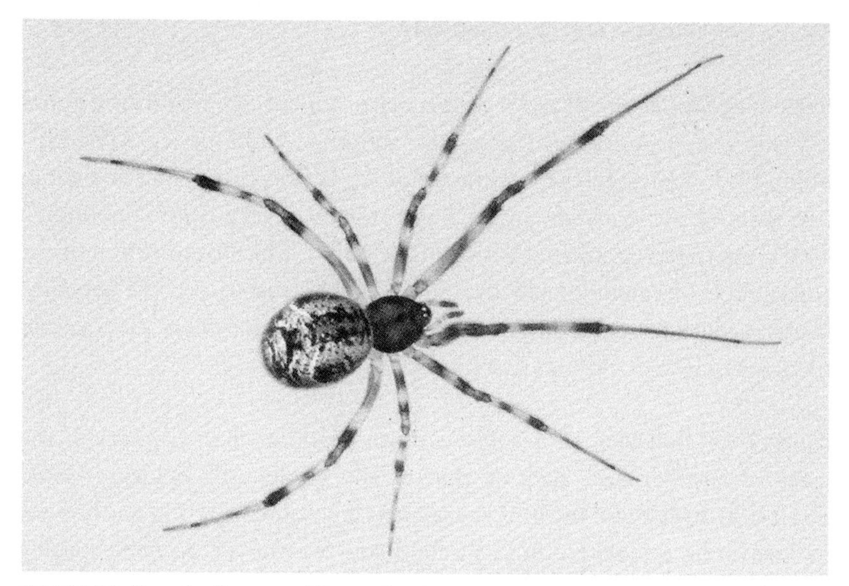

FIGURE 39. Female Common House Spider, *Achaearanea tepidariorum,* dorsal view.

Ecology and Behavior: Humans in the United States probably encounter this species more often than any other spider. It commonly lives in houses, basements and barns where it makes its web (the common "cobwebs") in corners of rooms and windows. Adults can be found in human habitations at all seasons of the year. They also may be encountered outside under stones and boards, on fences, and around bridges. The female usually hangs upside down in its tangled and highly irregular web. Flying insects that become entrapped in the web are injected with venom and mummified (encased in silk). The female suspends her egg sac in the web near where she rests. The egg sacs are paper-like and very tough, light brown in color, ovoid to pear-shaped, and from 6 to 9 mm in diameter. These spiders may live a year or more after they mature (Jackman, 1997). This species ranges throughout the U.S. and Canada.

Size: Length of female 5 to 6 mm; of male 3.8 to 4.7 mm (Kaston, 1978).

Family Theridiidae—*Argyrodes nephilae*

Identifying Characteristics: As with most other species within the genus *Argyrodes, A. nephilae* has a triangular abdomen and long legs. The tibia of leg I is longer than the metatarsus of leg I. This tiny spider is distinctive from other *Argyrodes* by having a mostly silvery colored abdomen which is sharply contrasted with a black venter. The silver portion of the abdomen is bisected dorsally by a mid-dorsal black stripe. The cephalothorax is dark brown as are the legs. Similar to *A. trigonum,* the head of the male of *A. nephilae* has two horns.

Ecology and Behavior: *A. nephilae* is often found at the periphery of the webs of large spiders such as that of the golden silk spider, *Nephila clavipes*. It lives there most of the time as a commensal. It is too tiny for its host to be interested in it as a food source. And, *A. nephilae* does its host no harm. However, tiny insects often get trapped in the host's web and may be devoured by *A. nephilae.* Thus, some might argue that

FIGURE 40. Female *Argyrodes nephilae* exhibiting her unique silver and black abdomen.

A. nephilae spends part of its existence in its host web as a kleptoparasite, stealing the food of its host. However, "kleptoparasite" is most likely not an accurate description as this tiny spider does no harm to its larger host and the food that it "steals" is too tiny for the host to eat. *A. nephilae* can also live an independent life and can spin a *Theridion*-like web among the branches of shrubs (Comstock, 1912).

Size: Length of female 3 to 4.5 mm; of male 2 to 3 mm.

Family Theridiidae—*Argyrodes trigona*

Identifying Characteristics: As its specific name implies, this small theridiid spider is recognized by its light brown triangular-shaped abdomen. The female is the larger of the sexes and the abdominal triangle contains the spinnerets at one angle and the pedicel at another. The third angle of the triangle is drawn out to a tip of the abdomen, which may vary in length and curvature from individual to individual. The cephalothorax and legs are slightly darker than the abdomen. On the cephalothorax the eyes are raised slightly. The male is smaller that the female and is usually slightly darker in color. The abdomen is triangular but never as large or as angular as the female. On the head of the male are two prominent horns tipped with tufts of hairs. The eyes are located at the base of the more posterior horn.

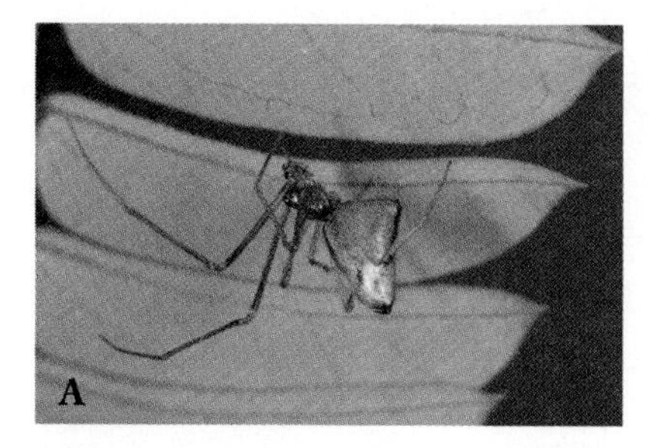

FIGURE 41. Female (left) and male (right) *A. trigona* recovered from irregular webs among fern leaves.

Ecology and Behavior: As with other spiders in this genus, *A. trigona* builds its web among larger orb webs where it lives a commensal life feeding on the smaller insect discards of the host spider. This species is also known to live independently in their own webs. Their snare is a very loose irregular web found beneath the leaves of low shrubs or weeds. The egg sac has the appearance of a small vase or acorn and is suspended in the female's web during mid-summer. The young emerge in late summer or early fall. *A. trigona* ranges throughout the eastern U.S. and Canada.

Size: Length of female 2 to 3 mm; of male 1 to 2 mm (Comstock, 1912).

Family Theridiidae—*Euryopis limbata* (Walckenaer)

Identifying Characteristics: This tiny spider is easily recognized by its uniquely shaped and distinctively colored abdomen, which is dorsoventrally flattened, somewhat heart-shaped, and pointed from behind. The

FIGURE 42. Female *Euryopsis limbata,* dorsal view

dorsum of the abdomen is largely black in color except for a contrasting silvery white stripe which forms a border around its sides and extends along the posterior half. The triangular shaped black portion of the abdomen is broken by two white spots anteriorly. The cephalothorax is also distinctly shaped, being as wide as it is long with its sides rounded. That part of the cephalothorax that extends beneath the abdomen is yellow while the anterior portion is black. The legs are yellow with darkened rings surrounding the ends of the joints. Leg IV is the longest.

Ecology and Behavior: This small spider is often found in shrubs where it probably stalks its prey rather than spinning a web to entrap them. It may even be found on the ground beneath stones or amongst mosses and lichens and leaf litter. They are very fast runners. Some authors report that this spider feeds on ants (Levi et al., 1990). During the summer, this spider may be collected by placing a white cloth or sheet beneath shrubs and shaking the shrubs in order to dislodge the spiders onto the sheets below. In the winter time, it may be found in leaf litter on the ground. The photographed specimen was actually collected in a crowded office within a university building where it was slowly descending through mid-air while attached to a silken dragline anchored to the ceiling. These peculiar spiders are found in the southern portion of eastern Canada and throughout most of the eastern one-half of the U.S. (Kaston, 1978).

Size: Length of female 3 to 4 mm; of male 3 mm (Kaston, 1978).

Family Theridiidae—*Latrodectus geometricus* (C. L. Koch)—Brown Widow

Identifying Characteristics: This small "widow spider" is easily confused with the similarly sized and colored common house spider, *Achaearanea tepidariorum. L. geometricus* is usually brown, but it may be grayish or black in color. As with most "widow spiders", the abdomen is rounded and has the familiar characteristic markings of an hourglass on the underside. However, in this species, the hourglass is not bright red as in

FIGURE 43. A female brown widow, *Latrodectus geometricus*.

the black widow, *L. mactans,* but is a washed out orange or pink. Dorsally, the abdomen has three white spots; posteriorly, the abdomen has a medial white band; and laterally, the abdomen has four diagonal white bands. The cephalothorax is dark brown and the light colored legs are banded in brown at the distal ends of each segment.

Ecology and Behavior: The habitat of *L. geometricus* is relatively similar to that of other widow spiders. It prefers dry undisturbed areas such as abandoned buildings and construction lots with discarded construction materials, cans, lumber, logs, and other debris. However, this species is more likely to be associated with human dwellings than *L. mactans*. It is more cosmopolitan in tropical climates, being found in habitats more similar to those of the common house spider, *Achaearanea tepidariorum.* The photographed female specimen was collected from beneath discarded construction materials from a developing resort area of Jacksonville, Florida. She guarded her spherical, fluffy egg sac (tufted with silk) which contained active spiderlings. The egg sac shape is unique for this species. Jacksonville, FL is near the northern limit of this spider's range. **This spider is venomous to humans** and should be approached with caution.

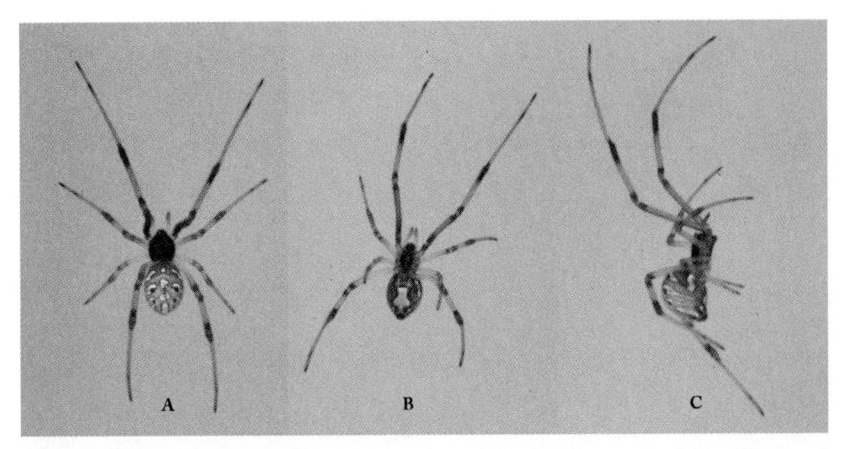

FIGURE 44 A-C. A female brown widow, *Latrodectus geometricus,* showing (**A**) dorsal; (**B**) ventral; and (**C**) lateral views.

Size: Length of female 5 to 6 mm; of male 3.4 to 4.5 mm (Kaston, 1978).

Family Theridiidae—*Latrodectus mactans* (Fabricius)—Southern Black Widow

Identifying Characteristics: The female is readily recognized by its shiny black body and legs. Its globose abdomen is decorated on its underside with a distinctive red hourglass-shaped marking. The red hourglass seems to be the most constant marking on this spider even though it can exhibit some intrapopulational variations. Some females may have the hourglass separated into two parts, but in most specimens it is complete. On most older females, the dorsum of the abdomen is solid black. In younger females, a row of red spots may extend down the mid-dorsum of the black abdomen. Many specimens have one or more red spots near the spinnerets. The closely related northern black widow, *Lactrodectus variolus,* nearly always has the red row of spots on the back and its red hourglass is almost always separated into two parts (see Figure 47). The much smaller male of *L. mactans* is beautifully and very conspicuously colored. It, like the female, has the hourglass marking on its abdomen

FIGURE 45. Adult female Southern Black Widow, *Latrodectus mactans,* exhibiting the distinctive red "hourglass" pattern on the underside of her abdomen.

although it is more of a red-orange color than red. Unlike the female, it displays a continuous or broken red-orange stripe bordered by white down the dorsal midline of the abdomen, as well as, four pairs of white stripes along the sides of the abdomen. Interestingly, young and subadult females are often marked very similar to the males.

Black Widow Bites: **This spider is venomous** and its capture should not be attempted except by professionally trained biologists. While the black widow spider may be encountered in any state except Alaska, it is particularly common in the southeastern U.S. and most reported human fatalities from its bite have occurred in this region. Most people are bitten when the spider is accidentally trapped against a portion of the body or when the web is accidentally touched. The bite of the female contains a powerful neurotoxin. Two red puncture marks will often be seen at the site of the bite and there is often a dull numbing pain around this area which may persist for up to 48 hours. This is accompanied by severe abdominal pain

FIGURE 46 A-B. Immature male (**A**) and immature female (**B**) southern black widows.

and a boardlike, rigid abdomen, along with rigidity and spasm of all large muscle groups. Tightness in the chest and pain upon breathing is common. Convulsions, paralysis and shock may occur. It has been reported that four to five percent of the untreated cases result in death. **A person bitten by this spider should see a physician immediately.** Medical treatment may include a dosage of antivenin, calcium gluconate (to control muscle pain), muscle relaxants, and treatment for shock. To avoid being bitten, one should stay away from places where these spiders live. If these spiders live near your house or in your basement, frequent cleaning should be done in order to remove the spiders and webs to decrease the possibility of accidental bites. Wear gloves and a longsleeved shirt while cleaning. If the area is infested, a good pest control company may be contacted to rid your property of these dangerous spiders.

Ecology and Behavior: The southern black widow makes an irregular web of very strong and coarse silk most commonly near the ground around tree stumps, woodpiles, under stones and loose bark, around water faucets, in holes in the ground, garages, barns, storage buildings and outhouses. It is a rather shy spider and will usually retreat into a hole or crevice when confronted by humans. Often the egg sac will be deposited within the irregular web and the female will guard it. The

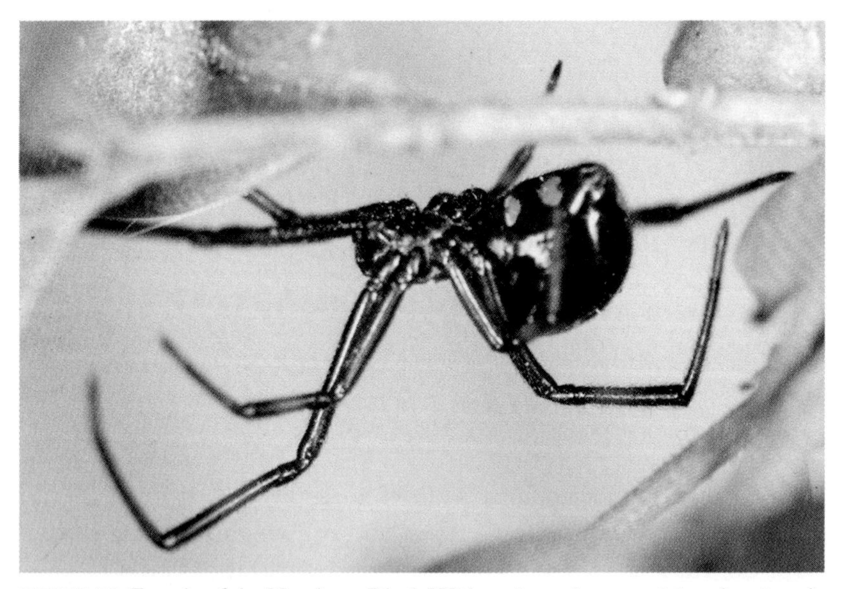

FIGURE 47. Female of the Northern Black Widow, *Latrodectus variolus,* showing the broken hourglass pattern on the ventral side of her abdomen.

nearly spherical egg sacs are characterized by being white to beige to light brown in color, about 10 to 12 mm in diameter, and pear shaped or with a distinctive nipple-like protrusion at the top. This spider is common throughout the eastern U.S. westward to Texas, Oklahoma and Kansas. It is most common throughout the southern states.

Size: Length of female 8 to 10 mm; of males 3.2 to 4 mm (Kaston, 1978).

Family Theridiidae—*Rhomphaea fictilium* (Hentz)

Identifying Characteristics: This light yellow to silvery-white spider is identified by the elongated abdomen and legs. The abdomen extends beyond the spinnerets a length longer than that of the body in front of the spinnerets. Leg I is the longest; however, each leg is longer than the entire length of the body, including the long abdomen. The cephalothorax has

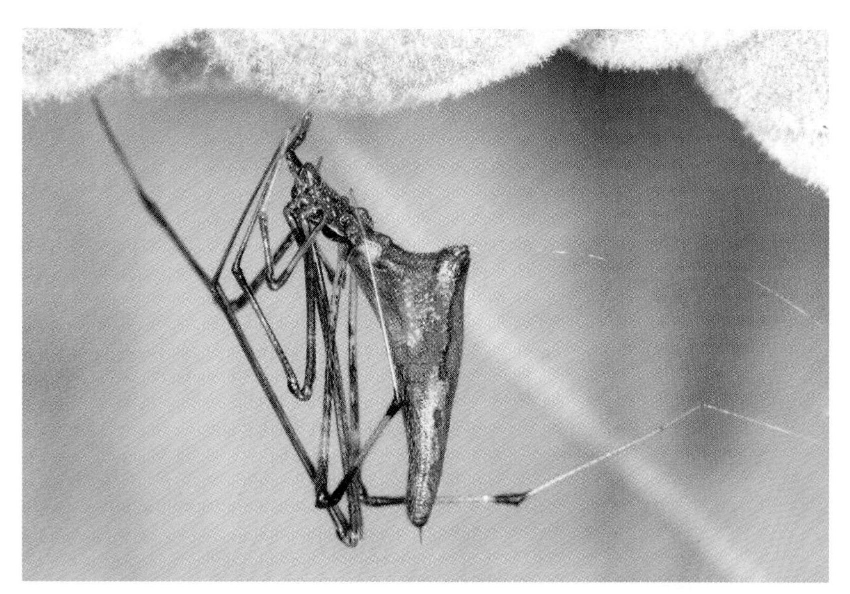

FIGURE 48. A female *Rhomphaea fictilium* displays her elongated abdomen.

three dorsal bands. The abdomen has three prominent longitudinal bands, one medial and two lateral.

Ecology and Behavior: *R. fictilium* is not a common spider but can be collected from bushes and grass. It constructs an irregular web like spiders of the genus *Theridion*. In its web the spider hangs motionless upside-down with its long legs pulled around its body. With close examination of a living specimen, the unusually long abdomen is quite limber. The spider can twist, turn and elongate its abdomen simulating an insect larva. Comstock (1912) hinted that this behavior may be a lure to attract insect prey. *R. fictilium* ranges throughout the United States.

Size: Length of female 5 to 10.5 mm; of male 4 to 7 mm (Kaston, 1978).

Family Theridiidae—*Spintharus flavidus* (Hentz)

Identifying Characteristics: This is one of our most colorful spiders. The cephalothorax is pale yellow. Most of the middorsum of the abdomen is a deep, bright yellow except for two pairs of light spots. White lines encircle the sides of the abdomen. Just inside the white lines are thin, highly contrasting, contiguous red and black lines. The sides of the angular shaped abdomen flare outward from their origin at the pedicel and become widest at a point about one-third the distance to the abdominal end. From this widest point, the sides taper sharply toward the posterior and terminate as a blunt point just over the spinnerets. The legs are pale yellow except a pair of red annulations at the distal ends of the tibia on legs I and IV. Legs I and IV are the same length and about twice as long as the second pair. Relative to body size, the male has longer legs and a more slender abdomen than the female. The lateral eyes on each side are contiguous. The posterior median eyes are widely spaced being separated by a distance equal to three to four times the diameter of one of the eyes.

FIGURE 49. Female *Spintharus flavidus* on a leaf.

Ecology and Behavior: This is a relatively common spider found on the under surfaces of leaves of low bushes. It might appear that the spider is resting on the leaf but a close examination shows that each foot is actually supported by a silken thread. This spider is most easily collected by using a sweep net in the understory bushes in shaded woodlands. This species is distributed from the eastern states to Oklahoma and Texas (Kaston, 1978).

Size: Length of female 4 to 4.5 mm; of male 2.8 mm (Kaston, 1978).

Family Theridiidae—*Steatoda grossa* (C. L. Koch)—False Black Widow

Identifying Characteristics: These spiders are dark in color and often mistaken for the Southern Black Widow, *Latrodectus mactans*. The abdomen, which may appear black in poor lighting conditions, is in reality

FIGURE 50. Female False Black Widow in her web. She was photographed in the interior of a drainage culvert.

a purplish to brownish purple, with a thin, light yellow to white line following the contour of the anterior portion of the abdomen. Behind this semicircular line, and extending down the middorsum, are three light yellow to white spots. In older females, these light markings may be faint or absent, giving a dark purplish-brown to black coloration to the abdomen. Even with the superficial resemblance to the black widow, this spider can usually be distinguished from that species by having the abdomen shaped more oval and not as globose. Additionally, it lacks the red hourglass mark of the black widow.

Ecology and Behavior: We have found this spider living together with the triangulate comb-foot spider, *S. triangulosum,* beneath manhole covers and within the manholes themselves. We have even found these spiders active during the winter months in such protected situations. This species also makes its snares under bridges, in culverts, and under rocks.

It is also common around human dwellings in much the same habitats as the house spider, *Achaearanea tepidariorum*. According to Kaston (1978: 115), it is one of the commonest "house spiders" of southern California. Kaston (1978) also cited a report of this species preying upon black widow spiders. This is a long-lived species, attaining an age of six years.

Size: Females 5.9 to 10.5 mm in length; males 4.0 to 7.2 mm (Kaston, 1978).

Family Theridiidae—*Steatoda triangulosa* (Walckenaer)—Triangulate Comb-Foot

Identifying Characteristics: The carapace is orange-brown. The ground color of the abdomen is yellow with two lateral ziz-zag brownish to purple stripes along the dorsum leaving a central band of yellow triangles stacked on top of one another. The legs are light-yellow, with slightly

FIGURE 51. Female *Steatoda triangulosa*.

darker rings at the ends of the joints. Through a dissecting microscope or a good hand lens, it may be seen that the hairs over the body all have thickened brown spots at their bases.

Ecology and Behavior: As with most theridiids, this spider spins an irregular web although it is unique in its design from that of other theridiids. Kaston (1978) gives a drawing of the characteristic *Stetoda* web. The web is constructed in dark corners of old buildings, outhouses, beneath large rocks, bridges, culverts and inside water meters. According to Moulder (1992), it has been reported to feed on ants. Mating occurs throughout the summer months and females may produce several spherical, white egg sacs which are suspended in the web. Some adult females may overwinter when winters are mild or when their snares are built in a protected and enclosed area. We have found this species occupying the same areas as *S. grossa*. This spider occurs throughout the U.S. (Kaston, 1978).

Size: Length of females 3.6 to 5.9 mm in length; of males 3.5 to 4.7 (Kaston, 1978).

Family Theridiidae—*Theridion frondeum* (Hentz)

Identifying Characteristics: *Theridion frondeum* is a moderate-sized theridiid spider with very distinctive coloration. Its background color is white but often with a tinge of yellow or green. Along the carapace and abdomen are median black markings that exhibit considerable variation. Some individuals may lack the black markings while others may have very broad dark bands on either or both the cephalothorax and abdomen. The photographed specimen, from Alabama, is intermediate in this feature with a moderately wide black band on the carapace and a median abdominal band reduced to two longitudinal stripes of triangular spots. Comstock (1912) has claimed that if one had twelve specimens from the same population in front of them, that the variablity in coloration would be so great as to lead one to think that there were at least

FIGURE 52. A *Theridion frondeum* female guarding her egg sac.

six different species! The legs are long, especially leg I. In this leg, the patella plus tibia is nearly twice as long as the carapace. The legs are a dull greenish to white color with distinctive black bands which delineate the distal ends of each segment in leg I, and in all but the femurs of legs II and IV, but not at all in leg III.

Ecology and Behavior: *T. frondeum* is one of the most common black and yellow-to-white theridiid spiders found in the litter of forests and fields. The photographed specimen was collected from leaves beneath blueberry bushes. Webs are spun inside folded leaves beneath tall grasses and shrubs. During July, the female spider deposits into the web a spherical, silky-white egg sac that is as large as the spider. The female guards the egg sac and remains with her young spiderlings after they emerge during August. This highly variable species occurs from southern Canada and New England south to Alabama and west to North Dakota (Kaston, 1978).

Size: Length of female 3 to 4.2 mm; of male 3 to 3.5 mm (Kaston, 1978).

Family Theridiidae—*Theridion glaucescens* (Becker)

Identifying Characteristics: This small round-bodied theridiid has an orange-brown carapace with a dark median V-shaped mark, broadening from the thoracic groove into the ocular region. The abdomen is a dirty white, dusted with grays and reds. Along the abdominal midline is a series of white, overlapping diamond-shaped spots arranged as a serrated stripe. Anteriorly, the abdominal white stripe has a dark spot. Laterally, the margins of the median white stripe are highlighted by black shading. The legs are darker than the basal color of the abdomen and have faint annuli. The male has exceptionally large palpal organs and the abdominal markings are similar to those of the the female, only sharper in color. *T. glaucescens* is easily confused with two very similar theridiid species common to the eastern U.S.: *T. differens* and *T. murarium.* All three species are similar in size, shape and markings. Positive identification of these three species is based upon the palpal organs in the male and the epigynum of the female (Emerton, 1961).

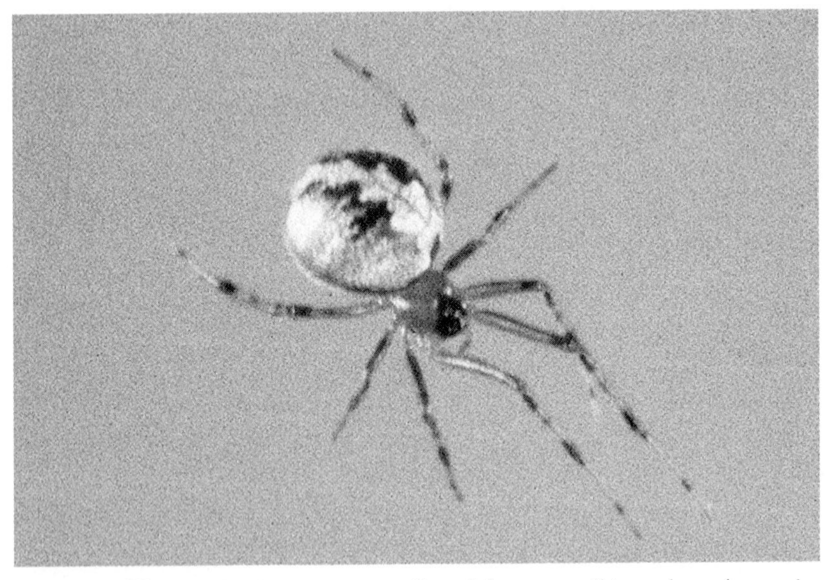

FIGURE 53. This tiny *T. glaucescens* was collected from a small irregular web spun in a folded leaf of a shrub.

Ecology and Behavior: Emerton (1961) described the webs of *T. differens* which are very similar to the webs of *T. glaucescens*. These irregular webs are usually spun in small trees or bushes three feet or less from the ground. The webs usually entangle several twigs and leaves. We have found webs of *T. glaucescens* around buildings, as well. The spider is often found in a retreat within the web that is composed of folded dried leaves fortified with webbing. Egg sacs are deposited in the retreat with the spider. This species is distributed from southern Canada and New England south to Florida and west to Texas and Nebraska.

Size: Length of female 2.2 to 3.0 mm; of male 1.6 to 2.5 mm (Kaston, 1978).

Family Theridiidae—*Theridula emertoni* (Levi)

Identifying Characteristics: *T. emertoni* is an extremely tiny spider and one that would be easily overlooked were it not for its brilliant colors. In the female, the legs and margins of the cephalothorax are pale yellowish to amber. The central portion of the cephalothorax is black. The abdomen is brilliant reddish-orange except for a large middorsal yellow spot and two lateral black spots. The abdomen is unusually wide and has lateral protuberances overlain by the black spots. In the male, the abdomen is not as distinctively colored as in the female and the lateral humps are narrower. This color description differs from that given by Kaston (1978) where he states, "The abdomen is greenish gray with a white spot in the middle, a large black spot around the spinnerets and one on each side where the abdomen is drawn out into a hump."

Ecology and Behavior: These spiders are most easily collected using sweep nets drawn through grasses and small bushes. Kaston (1978) reported that he collected them in bushes and hemlock trees. Records of their web are lacking. We observed one specimen dangling from a silken gossamer thread as they seized and captured prey. Comstock (1912) reported

FIGURE 54. A female *T. emertoni* displays her distinctive red, black and yellow colors.

that the females carry their egg-sacs attached to the spinnerets like the wolf spiders, thus decreasing the necessity of a retreat. They range from New England and southern Canada south to Alabama and west to Wisconsin.

Size: Length of female 2.3 to 2.8 mm; of male 2 to 2.3 mm (Kaston, 1978).

Family Theridiidae—*Tidarren sisyphoides* (Walckenaer)

Identifying Characteristics: The ground color of the dorsum varies from a light brownish tan to a grayish black. The legs are spotted and banded. The abdomen is usually with a pattern of spots on the anterior portion. The most distinctive color feature is a narrow middorsal longitudinal white line extending from the highest point of the abdomen to the anal

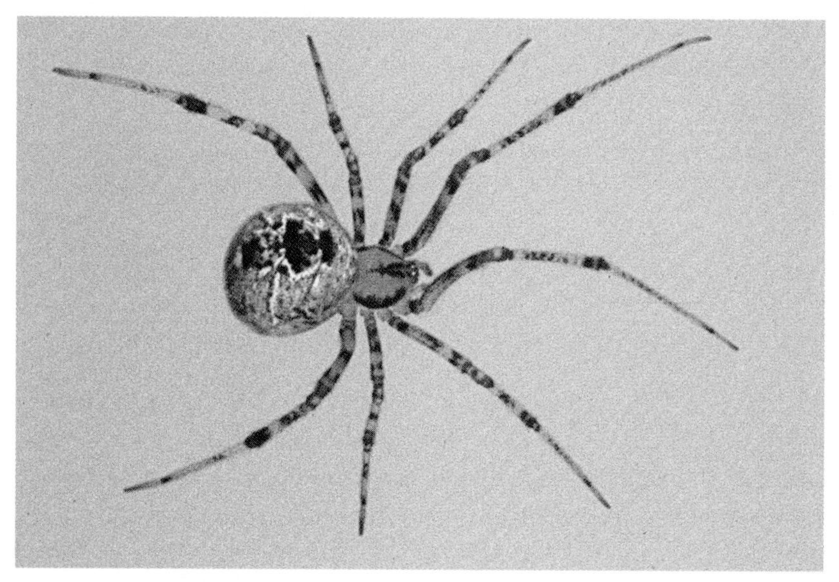

FIGURE 55. Female *Tidarren sisyphoides* collected from retreats constructed of a rolled leaf within irregular webs.

tubercle. The epigynum has a cone-shaped knob and from the side it appears beaked. The males are very small, being only about one-third the length of the females. The adult male is distinctive in that it has only a single pedipalp which is disproportionately large.

Ecology and Behavior: According to Levi (1990), the female lives in a curled dead leaf hanging in the web. The tiny male hangs nearby. The male amputates one of his enormous palpi before his final molting. We have found this species making its irregular web in the corners of outdoor porches and patios and in the eaves of houses. All webs have at least one dead leaf and other debris brought into the web to use as a hiding place for the female, and in which to place her egg sac. During the daylight hours, the adult female may be found hiding inside the leaf. In the southern states, this spider may live in essentially the same type of habitat as the common house spider, *Achaearanea tepidariorum,* and the two species often construct very similar webs which may be placed side-by-side. This spider is found throughout the southeastern U.S. westward to California.

Size: Length of female 5.8 to 8.6 mm; of male 1.2 to 1.4 mm (Kaston, 1978).

FAMILY LINYPHIIDAE—Sheetweb and Dwarf Weavers

According to Roth (1993), this is the largest family of Araneae in the world, with about 162 genera and 869 described species, with many more undescribed species in North America. It has been the conventional taxonomic treatment to divide this family into two subfamilies: the Linyphiinae (sheetweb weavers) and Erigoninae (dwarf weavers). Some taxonomists feel that there are many reasons why this division into

FIGURE 56. The "bowl and doily" web of the linyphiid spider, *Frontinella pyramitela*.

only two subfamilies is unsatisfactory. Millidge (1980) has actually divided this family into multiple subfamilies, but his work has not been used regularly by most systematists. A useful privately published paper on the Family Linyphiidae is available through the American Arachnological Society (Roth and Buckle, 1986).

Many beginning students of spider identification may have problems distinguishing some members of this family from those of the Theridiidae. However, the following distinguishing features may help: The linyphiids lack the comb of serrated bristles on the tarsus of Leg IV, and usually have thinner legs which usually bear spines. The linyphiids have more robust chelicerae and their margins are always provided with stridulating ridges; the linyphiid labium is rebordered (edge is thickened); and, a colulus is present. Finally, linyphiids are usually very small, most being less than 3 mm in length. Linyphiids build a flat or irregular web. Linyphiids and theridiids may also be distinguished on the basis of genitalia differences, which is beyond the scope of this book.

Family Linyphiidae—*Florinda coccinea* (Hentz)

Identifying Characteristics: The color of this tiny spider is crimson, red or orange, except for a distinctive black area around the eyes, the last three segments of the pedipalps, and the tip of the tubercle at the end of the abdomen. Besides the distinctive coloration, this spider has a pronounced tubercle at the end of the abdomen that projects out over the spinnerets.

Ecology and Behavior: This tiny spider builds its sheet web in grasses and low bushes. We have found both male and female in their webs during the late summer and early fall. This does not seem to us to be a particularly common species. Its web resembles that of the Bowl and Doily Spider, *Frontinella pyramitela*. We have found an occasional *Florinda* web among numerous webs of the Bowl and Doily. In Alabama, for every *Florinda* web we find, we encounter approximately 100 Bowl and Doily

FIGURE 57. Female *Florinda coccinea,* lateral view.

webs. When the web is approached by a human collector, and disturbed slightly, the male usually will drop out of the web onto the ground. With additional disturbance, this action is usually followed by the female. This spider usually builds its web over much ground litter and thick plants. When it falls from its web, it almost assuredly escapes among the ground cover. Comstock (1912) reported collecting this species only with a sweep net. In Florida, it may be commonly found on dew-covered lawns during early mornings among hundreds of *Florinda* webs. The species is more common in the south, being distributed from Maryland west to Illinois and south to Florida and Texas (Kaston, 1978).

Size: Length of female 3.5 mm; of male 3 mm (Kaston, 1978).

Family Linyphiidae—*Frontinella pyramitela* (Walckenaer)—Bowl and Doily Spider

Identifying Characteristics: In the females, the carapace is homogeneously brown in color. The abdomen has a broad median chocolate colored stripe bordered by a white stripe on each side. The white stripes usually have scalloped edges. The venter is mostly chocolate colored. When viewed from the side the abdomen is about as high over the spinnerets as it is at its anterior end. The legs are pale cream in color, sometimes with a pale greenish tinge. The males, which are smaller in size, have their carapace and legs tan in color. The abdomen is a dull white, becoming brownish at the posterior end. The venter is chocolate in color.

Ecology and Behavior: This spider takes its common name from its very characteristic web. The web consists of two portions: an upper, bowl-shaped sheet which sits suspended a few millimeters above a flat,

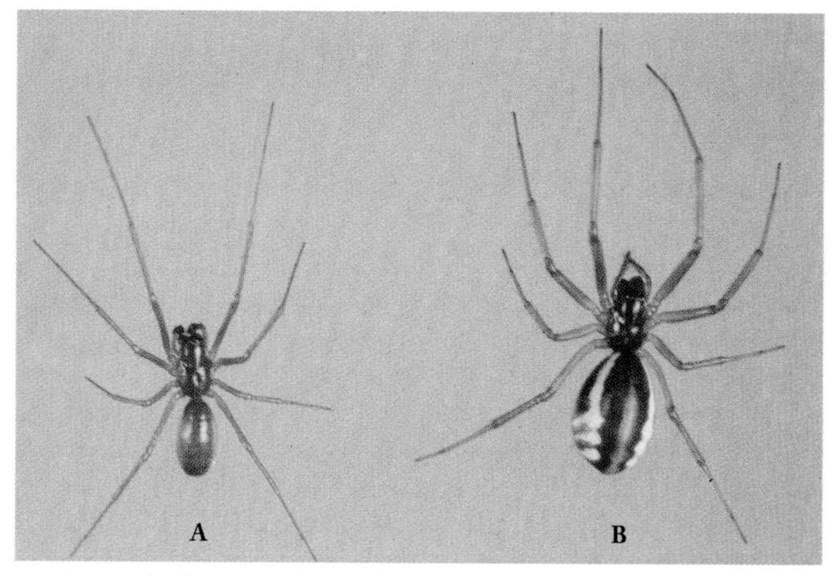

FIGURE 58 A-B. The Bowl and Doily Spider, *Frontinella pyramitela*. (**A**) Dorsal view of male; (**B**) dorsal view of female.

horizontal sheet reminiscent of a "doily". It could just as aptly been named the "cup and saucer" spider. A large number of irregular threads serve to anchor the bowl to plant materials above. The threads serve also to deflect flying insects into the bowl below. The spider, which hangs upside down beneath the bowl, bites the insect through the webbing of the bowl and pulls it through. Leafhoppers form a large portion of this spider's diet (Fitch, 1963). This distinctive snare is usually constructed in low weeds and bushes mostly within a meter of the ground. The spider is equally at home in ornamental shrubbery as it is in natural woodlands. The spider matures in late spring. During the summer, both male and female may be seen occupying the same web. The spider persists throughout the summer and up until November in warmer areas of the eastern U.S. Occasionally, the adults will survive the winter if mild. During 1997–98, an unusually warm winter, we observed adult females still in their webs in southeastern U.S. woodlands in late January. This species is distributed from southeastern Canada, throughout the entire U.S., and southward to Central America. This spider is also known under the name *Frontinella communis* (Hentz).

Size: Length of female 3 to 4 mm; of male 3 to 3.3 mm (Kaston, 1978).

Family Linyphiidae—*Neriene radiata* (Walckenaer)—Filmy Dome Spider

Identifying Characteristics: This spider is also known as *Prolinyphia marginata* (C. L. Koch). This is one of our commonest and most handsome species of woodland spiders. However, because of its small size, its beauty is seldom appreciated by most persons. The middorsum of the cephalothorax is brown with the margins being outlined in broad white bands. A lateral view of the abdomen shows that it is flat on top and is highest posteriorly. The abdomen has a rather complex, but distinctive color pattern. The colors are pale yellow and purplish brown. There is a wide, broken middorsal dark stripe that extends from the anteriormost part of the abdomen nearly to the spinnerets. Bordering this are two

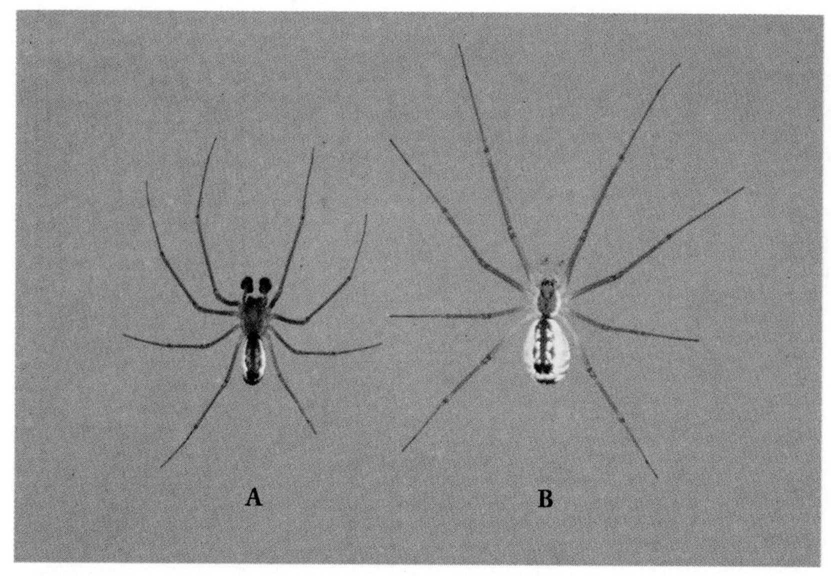

FIGURE 59. The Filmy Dome Spider, *Neriene radiata.* Dorsal view of male (**A**); and female (**B**).

wide lateral bands of yellow or white. Anteriorly, two transverse yellowish to white bands extend across the dark middorsal stripe and almost, but not quite, bisect it into three parts. Behind this, another wider transverse yellow band does cross this middorsal stripe, breaking the stripe into a large dark spot just above the spinnerets. Along the sides, there are alternating dark and light stripes, the anterior ones being oriented lengthwise while the posterior ones are arranged vertically. The venter is dark brown.

Ecology and Behavior: The common name of this spider comes from its characteristic web which is in the form of a dome about four or five inches in diameter and which is supported by a maze of silk threads. The web is placed on low shrubs, small trees, rock outcrops and stone walls, usually within and along the edge of deciduous woodlands. The spider positions itself upside down underneath the top of the dome. Flying insects are

FIGURE 60. Domed web of *Neriene radiata* showing the female suspended just beneath the apex of the dome.

knocked from the air as they fly unknowingly into the supporting maze of threads above the dome. They then drop onto the dome where the spider seizes the insect by biting through the dome webbing. It then cuts the web, pulls the insect through, mummifies it, and later repairs the cut in the dome. Both male and female are usually seen residing in the same web. Mating and egg laying take place throughout the summer. Young spiders may be seen in their small webs during August and September.

Size: Length of female 4 to 6.5 mm; of male 3.4 to 5.3 mm (Kaston, 1978).

FAMILY TETRAGNATHIDAE—Longjawed Orbweavers

The spiders of this family have been considered by some authors to represent only a subfamily of the Family Araneidae. However, most authorities feel that the following combination of characters set them apart enough

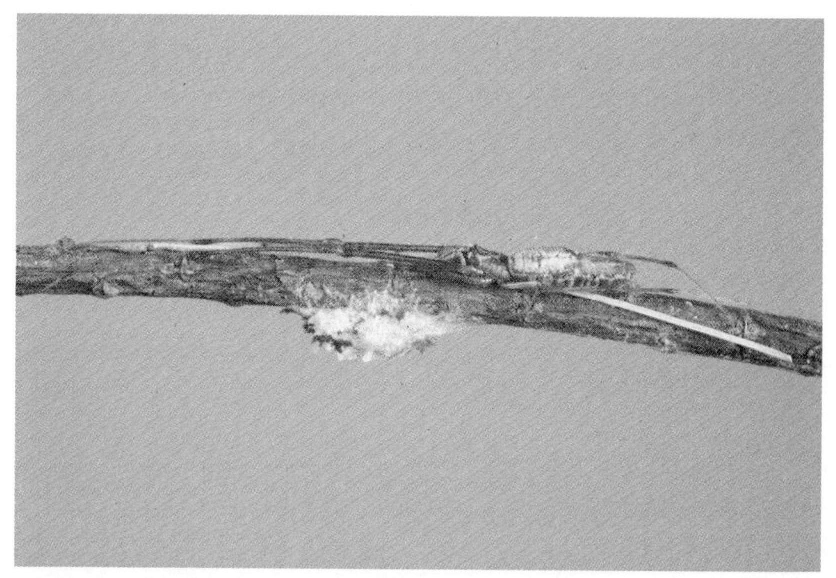

FIGURE 61. A female Longjawed Orbweaver guards her egg sac. The tetragnathids typically camouflage themselves by stretching out along a look-alike twig.

to warrant placement into their own family: a rebordered labium; cheliceral margins with stout teeth; colulus; enlarged chelicerae, especially in males of some genera; base of femur with either short or long trichobothria; epigastric furrow strongly procurved; rudimentary boss on chelicerae; and, femora I and II usually extended forward while at rest. Most tetragnathids construct horizontal or oblique orb webs with an open hub.

This family consists of three subfamilies: Metinae containing the genus *Leucauge,* Nephilinae containing the genus *Nephila,* and Tetragnathinae containing the genera *Glenognatha, Pachygnatha* and *Tetragnatha.* Nephilinae is tropical or subtropical but the other members of the family are distributed throughout the eastern U.S.

Family Tetragnathidae—*Leucauge venusta* (Walckenaer)—Orchard Spider

Identifying Characteristics: This is a beautiful, multicolored spider. The carapace is yellowish gray to green with a thin dark line down the center

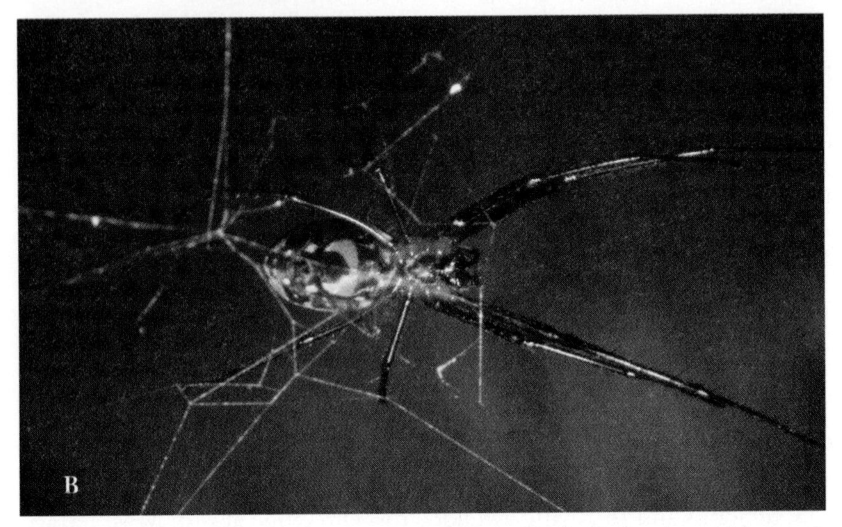

FIGURE 62 A-B. (**A**) Female *Leucauge venusta* displaying its distinctive multicolored abdomen; (**B**) the ventral abdomen of the same specimen showing the distinctive red spot which leads some to erroneously believe that this is a kind of "black widow".

and a pair of dark stripes on the sides. The sternum is green. As seen from above, the oval shaped-abdomen has a background color of silvery iridescence with a dark median band with lateral branches that extend backward. The sides and venter of the abdomen have silver maculations. The sides of the abdomen are yellowish with red spots near the posterior end. The ventral side of the abdomen has a distinctive red spot near the middle.

Ecology and Behavior: As implied by the name "Orchard Spider", this arachnid builds its horizontal web commonly in orchard trees. However, its main habitat is moist, open hardwood forests. Here the webs are found among the low limbs of small trees and shrubs. The chief prey consists of small flying insects. While waiting for its prey, the spider hangs upside down in the center of its web. If the spider is disturbed, it immediately drops to the ground in an attempt to escape. According to Jackman (1997), the egg sacs are "loose and fluffy, made of orange-white silk. They are 8–9 mm in diameter and hold several hundred eggs." This species ranges from New England south to Florida and west to Nebraska and Texas (Kaston, 1978).

Size: Length of female 5.5 to 7.5 mm; of male 3.5 to 4.0 mm (Kaston, 1978).

Family Tetragnathidae—*Nephila clavipes* (Linnaeus)—Golden Silk Orbweaver

Identifying Characteristics: The cephalothorax is black, but this dark color is obscured, except in spots, by silver-colored hairs which cover most of this region. The ground color of the abdomen is golden olive to yellowish brown with 5 or 6 pairs of yellow to white spots extending down the - dorsum. A conspicuous transverse white line extends across the anterior part of the abdomen. The abdomen is elongate and the sides are nearly parallel. One of the most distinctive identifying characteristics is the conspicuous tufts of hairs on the femora and tibia of legs I, II and IV. The female is a large spider, being up to 25 mm or more in length. The male is only one-fourth this size. Overall, the male is dark brown on both body and legs. Its legs lack the tufts of hairs found in the female.

Ecology and Behavior: This spider is widely distributed in the southeastern U.S., particularly along the Atlantic and Gulf coastlines. It is essentially a tropical to subtropical species and likely cannot overwinter

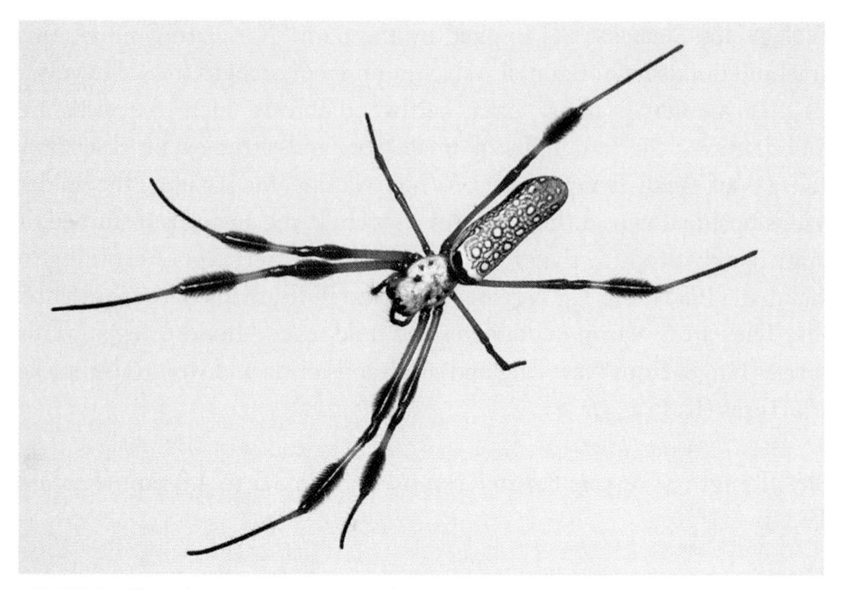

FIGURE 63. Female *Nephila clavipes* in her web of golden silk.

outside of these warmer, more humid states. It builds huge, slightly inclined webs up to a meter in diameter, usually in heavily shaded hardwood or palm forests. Its webs may be found from a meter to three meters off the ground. The viscid spiral of the web is unique in being golden-yellow in color, rather than white as in most spiders. The dry silk is white. The web of young spiders is usually a complete orb, but the adults build only the bottom part leaving the top of the orb incomplete. The web is further unique in that the radii are pulled out of their normal position and this gives a notched or looped appearance to the viscid lines. The spider hangs throughout the day on the lower side of the inclined web and feeds on flying insects.

Size: Length of females 18–25 mm; of males 4 to 8 mm (Kaston, 1978).

Family Tetragnathidae—*Pachygnatha tristriata* (C.L. Koch)—Thickjawed Orbweaver

Identifying Characteristics: At first glance, *Pachygnatha* appear to be either araneids with oversized chelicera or tetragnathids with short, rounded bodies. Actually, the somewhat atypical *Pachygnatha* are placed in the family Tetragnathidae for several reasons: (1) they have enlarged chelicera especially in the male; (2) they have long legs in which the front two pairs extend forward when at rest; and (3) they build orb webs in close proximity to water. *Pachygnatha* are unique among tetragnathids by having a much shorter body with a more rounded abdomen (length about 1.5 x width). *Pachygnatha* chelicerae are shorter and stouter than those of most tetragnathids; and, their long legs are spineless. Among the *Pachygnatha* the anterior and posterior lateral eyes are confluent while among other tetragnathids they are separated. *P. tristriata* has an

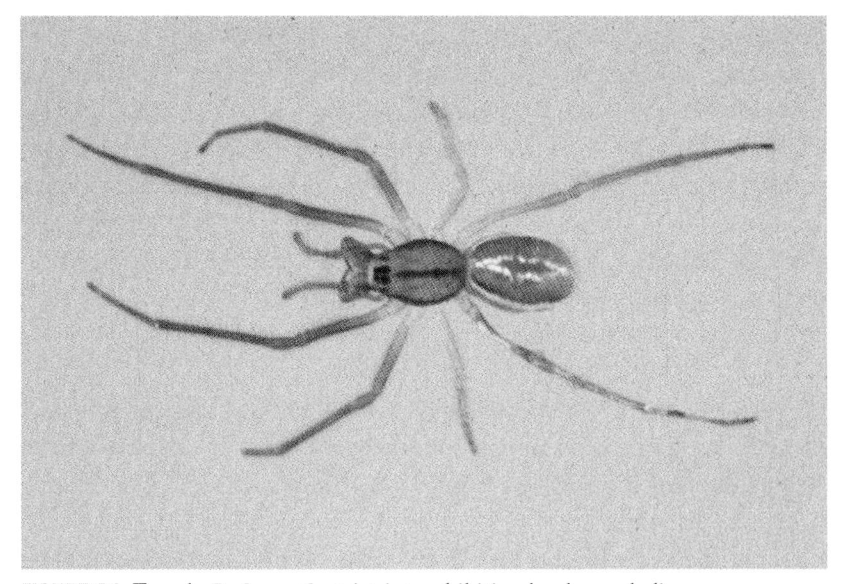

FIGURE 64. Female *Pachygnatha tristriata* exhibiting her large chelicerae.

overall yellowish brown color. The carapace is darker along its margins and has its highest point at a hillock upon which the posterior median eyes are situated. The yellow brown abdomen has a grayish folium pattern with dark borders.

Ecology and Behavior: *Pachygnatha* are commonly found in close proximity of water. The construction of orb webs has been attributed to only the immature spiders. Adults apparently do not construct webs but lurch among the vegetation in search of suitable prey. *P. tristriata* can be found throughout the eastern states. Its range extends from Florida northward into New England and eastern Canada and west to Texas and Nebraska.

Size: Length of female 5.5 to 6.5 mm; of male 5 to 5.5 mm (Kaston, 1978).

Family Tetragnathidae—*Tetragnatha elongata* (Walckenaer)—Elongate Stilt Orbweaver

Identifying Characteristics: The carapace is reddish brown with gray markings. The abdomen is dull silver with brownish pigment forming a central dark band with a pair of large lobes near the center of the abdomen. The abdomen of the female is widest near its junction with the thorax and tapers gradually toward the posterior end. The whole body is very slender and elongate. The first pair of legs is about ten times the length of the carapace. The chelicerae are sinuate, being almost as long as the carapace in the female, and longer in the male. The lateral eyes of each side are closer together than are the median eyes.

Ecology and Behavior: These spiders build an inclined orb web in grassy fields, meadows and bushes near lakes and streams. The female sits on the open hub of the web (Kaston, 1978). Adults are found from March through September. During mating, the male and female hold each other by the ends of the mandibles. The plano-convex egg sacs are

FIGURE 65. A female *Tetragnatha elongata* showing her enlarged chelicera.

attached to twigs and are covered with greenish silk threads. This species is found throughout the U.S., Canada and into Alaska.

Size: Length (exclusive of chelicerae) of female 9 mm; of male 7.5 mm (Kaston, 1978).

Family Tetragnathidae—*Tetragnatha laboriosa* (Hentz)—Silver Longjawed Orbweaver

Identifying Characteristics: The legs and carapace of *T. laboriosa* are amber brown in color. In comparison to other members of the genus, the chelicerae are short and held in a vertical position. In the male, the chelicerae are slightly larger than those of the female and are curved outward. Regardless of sex, the extended chelicerae are less than two-thirds the

FIGURE 66. A female *Tetragnatha laboriosa* sitting on a log near a stream.

length of the carapace. The abdomen is elongate and has a silvery sheen which is broken by an amber colored line in the middorsum.

Ecology and Behavior: Populations of *T. laboriosa* can become especially numerous and individuals may be collected in large numbers from shrubs and tall grasses with sweep nets. Otherwise, they are generally overlooked. Individuals with their elongate bodies cling to stems of vegetation and are nearly invisible. These spiders will construct complete orb webs in a more or less vertical plane which are not often noticed. They are most active at sundown, when they are likely to frequent their webs. When the spider is disturbed in its web, it will drop to the ground or water and scurry for shelter. *T. laboriosa* webs are common along small creeks and ponds, and are constructed wherever they can shelter their webs from strong wind. Based upon prey left in intact webs along streams, *T. laboriosa* feeds on small dipteran insects such as gnats, midges and mayflies. Most populations flourish between May and September and a single female will deposit 40 to 76 eggs in egg sacs secured to vegetation. *T. laboriosa* is common throughout the United States, Canada and Alaska.

Size: Length of female (exclusive of chelicerae) 6 mm; of male 5 mm (Kaston, 1978).

Family Tetragnathidae—*Tetragnatha straminea* (Emerton)

Identifying Characteristics: *Tetragnatha straminea* shares common features with two other tetragnathids (see Table 1). Similar to *T. versicolor*, the chelicerae are large and are more than half the length of the carapace. Similar to *T. laboriosa*, the abdomen of *T. straminea* is more elongate. The abdomen of *T. straminea* is more slender than other tetragnathids of the eastern U.S., being four to five times as long as wide. Different from any other tetragnathids, the posterior row of eyes is recurved so that the

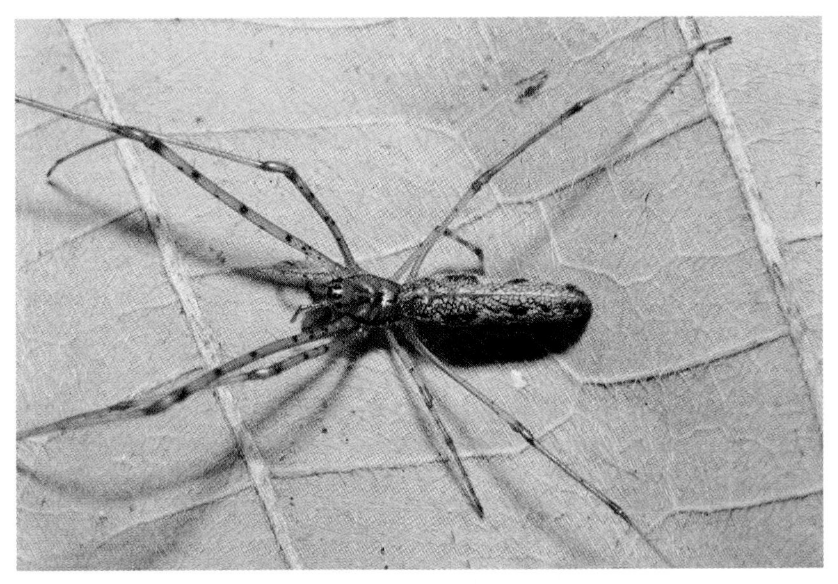

FIGURE 67. Female *Tetragnatha straminea.*

TABLE 1. Comparison of the Common Species of the Subfamily Tetragnathinae

SPECIES	LE = ME *	Chelicera Length**	Abdominal Length***	Abdominal Markings
Pachygnatha tristriata	LE confluent	<50% carapace	Length 1.5x width	Abdomen with a silver border
Tetragnatha laboriosa	LE=ME	≤50% carapace	Length 3x width	Abdomen mostly silver
Tetragnatha straminea	LE>ME	60–80% carapace	Length 4–5x width	Abdomen with a silver border
Tetragnatha versicolor	LE<ME	50–67% carapace	Length 2.5x width	Abdomen with a silver stripe
Tetragnatha elongata	LE<ME	100% carapace	Length 2.5x width	Abdomen is dull silver with brown markings
Tetragnatha viridis	LE=ME	≤50% carapace	Length 3x width	Abdomen is dull yellow with green markings

*LE=ME: Distance between anterior and posterior lateral eyes relative to anterior and posterior medial eyes.
**Chelicera length relative to carapace length
***Abdominal length relative to abdominal width.

lateral eyes are further apart than the median eyes. The only silver on the abdomen borders the folium.

Ecology and Behavior: *T. straminea* constructs orb webs between twigs of bushes and shrubs near streams and ponds. However, during the daytime, it is seldom found in its web but most often is seen lying lengthwise along the stems of plants where it is camouflaged. It ranges from New England, south to Florida and west to Minnesota and Texas (Kaston, 1978). However, it is more common in northern states (Comstock, 1912).

Size: Length of female 8 mm; of male 6.5 mm (Kaston, 1978).

Family Tetragnathidae—*Tetragnatha viridis* (Walckenaer)

Identifying Characteristics: This spider is easily recognized as a tetragnath because it exhibits the typical features of large and extended chelicera, an elongated abdomen, and long legs with numerous spines. This species is distinguished by its green color, which provides excellent camouflage among the green pine needles where it lives. Of the more commonly encountered tetragnathids, it most closely resembles *T. laboriosa* than the other species of the genus. The distance between the anterior and posterior lateral eyes and the anterior and posterior medial eyes are the same. The chelicerae are half the length of the carapace. The abdomen is about three times as long as wide.

FIGURE 68. A female *T. viridis* camouflaged among the green needles of a loblolly pine tree.

Ecology and Behavior: This spider was first collected in Georgia and described in 1841 (Chamberlin and Ivie, 1944). Since then, it has taken on other specific names such as *T. pinicola* and *T. pinea*. These synonyms account for the spider's tendency to inhabit trees and bushes of pine forests. They will construct small complete orb webs among the lower limbs of small pine trees. However, they do not occupy their webs but will attain a motionless position aligning its body along a green pine needle or twig with forelegs extended. This green tetragnath has not been included in most spider texts. It is limited in its range, being restricted to young southern pine forests throughout Alabama, Georgia and Florida during the warmer months. Its green color and motionlessness make it nearly impossible to see in pine foliage. Foremost, the enthusiast in search of tetragnath spiders, generally does not use a sweep net to collect in pine trees. *T. viridis* is most likely an overlooked spider rather than a rare spider.

Size: Length of female and males (excluding chelicerae) is 4 to 5.5 mm.

FAMILY ARANEIDAE—Orbweavers

The araneids are sedentary spiders that are easily recognized by their orb webs. For the sole purpose of snaring prey, most construct their orb webs on the vertical or near-vertical plane. Some species of araneids will sit in the center of their web, head downward and in constant vigilance of web vibrations signaling an entangled flying insect. Other araneids seek a peripheral hiding place in vegetation or a folded leaf, yet they still monitor web vibrations from silken leaders. Rarely do most araneids move, except to mummify a captured insect or to escape danger.

Orb webs are also constructed by a small number of spider families other than araneids. The spiders of the family Uloboridae construct orb webs which are especially small, tight webs spun on the horizontal plane. The orb webs of spiders of the family Tetragnathidae are araneid-like but are more loosely arranged with fewer radii and spiral strands. Tetragnathid orb webs are usually found around bodies of water.

FIGURE 69. A dew covered orb web of the araneid spider, *Cyclosa conica.* The tiny camouflaged spider is resting just below the center of the web.

Other than their aerial orb webs, the araneids are distinguished by having large chelicerae with a large boss or lateral condyle. The fang furrow is bordered with stout teeth. A colulus is present and the legs are very spiny.

Araneidae is a very large family, consisting of 31 genera and 155 species. We have given accounts of spider species representing five subfamilies: Argiopinae, containing the genera *Argiope* and *Gea,* Cyrtophorinae, containing the genus *Mecynogea,* and Gasteracanthinae containing the genera *Gasteracantha* and *Micrathena.* Nephilinae, containing the genus *Nephila,* and Metinae, containing the genus *Leucauge,* are two subfamilies that recently have been regrouped with the family Tetragnathidae by other authors (Roth, 1993).

Family Araneidae—*Acacesia hamata* (Hentz)

Identifying Characteristics: *Acacesia hamata* has a distinctive rhomboid shaped abdomen with a grayish ground color. The cardiac mark is brown and extends more than half the length of the abdomen. The cardiac mark is bordered by a broad band lighter in color. The thoracic groove on the carapace is longitudinal and extends forward to the cervical groove. The anterior and posterior rows of eyes are strongly recurved and are nearly vertical on the front of the head. The legs of the female have very few spines. The male resembles the female in color and markings. The legs of the male are more spiny, especially the tibia of leg II.

Ecology and Behavior: *A. hamata* constructs its web in open woods and shrubs three to four feet from the ground. *A. hamata* is strictly nocturnal, building its web after dark and removing it before sunrise. The spider is most likely found in its web between late evening and early morning

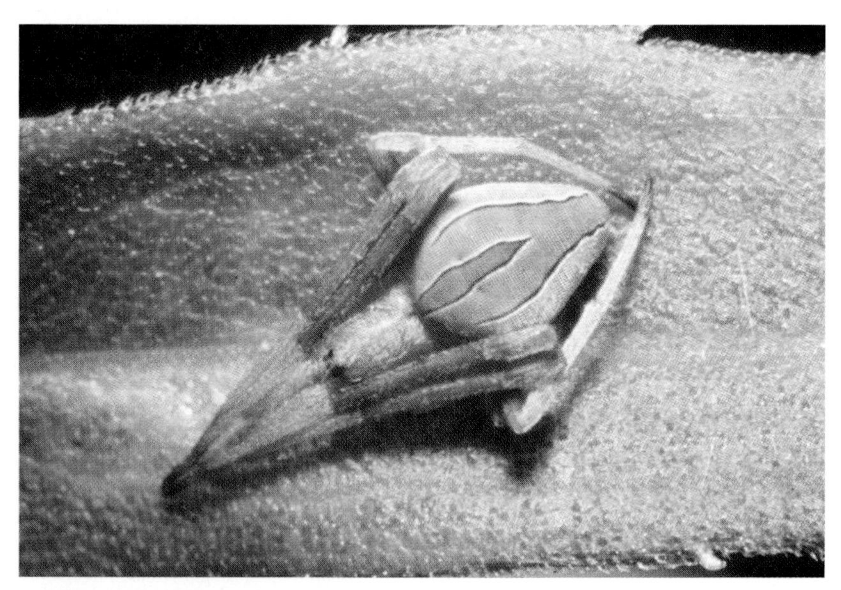

FIGURE 70. Female *Acacesia hamata*

during mid summer. The orb web is usually 20 to 25 cm in diameter and vertical. It is supported above by two nearly horizontal framing strands of silk and below by three diagonal framing strands. The web is tightly woven with about eighty closely spaced spiral strands. When startled with bright lights, females have been observed to completely consume their webs in less than a minute. At first, the spider drops and clips the lower framing supports of the web. With quick left to right sweeping movements across the web, she rolls up the web to the horizontal as though it were a window shade being rolled up. The spider then clips one of the horizontal supports at one side and climbs the opposite horizontal framing strand, consuming the rolled up webbing as she climbs. *A. hamata* is not a common spider, but when discovered dozens of webs are found in close proximity. It is found throughout the eastern U.S. westward to Texas and Illinois (Kaston, 1978).

Size: Length of female 4.7 to 9.1 mm; of male 3.6 to 5 mm (Kaston, 1978).

Family Araneidae—*Acanthepeira stellata* (Marx)—Starbellied Orbweaver

Identifying Characteristics: The abdomen contains several thick cone-like protuberances which form a star-like shape. These tubercles consist of a median anterior protuberance that overhangs the cephalothorax and bears a white triangular spot at its base. There is a median posterior protuberance, two lateral pairs on the lower regions, and three lateral pairs on the upper region of the abdomen. The abdomen is brown, and its projections are highlighted with a cream color. The carapace is reddish brown with black along the sides and is clothed with light colored downy hairs. The legs are yellow with brown rings.

Ecology and Behavior: *Acanthepeira stellata* builds its orb webs in the vertical position in open sunny areas, usually within four feet of the ground. The web is secured by foundation stands to slender stalks and leaves. The spider typically hangs in the hub of its web with its head pointing

FIGURE 71. Female *Acanthepeira stellata*. Note the distinctive cone-like abdominal tubercles.

toward the ground. If disturbed in the least, it quickly drops from its web and seeks a camouflaged, motionless position among the lower vegetation. *A. stellata* is more common in wooded areas, but it may be found in open meadows with tall grasses. The young spiderlings emerge in the fall, overwinter, and mature during the early summer months. In the spring, spiderlings leave the adult's web on long strands of gossamer that enable them to float on air currents for thousands of feet. *A. stellata* ranges from New England and Canada, south to Florida and west to Kansas and Arizona (Kaston, 1978).

Size: Length of female 7 to 15 mm; of male 5.1 to 8.1 mm (Kaston, 1978).

Family Araneidae—*Araneus bicentenarius* (McCook)

Identifying Characteristics: This beautiful spider was collected by McCook during 1882 at the bicentennial of the city of Philadelphia, hence the species name "*bicentenarius.*" The carapace is dark brown. The abdominal dorsum is bright green with a distinctive folium created by two parallel zig-zag brown lines beginning a short distance behind the pedicel area and extending posteriorly and converging at the spinnerets. The abdominal area just behind the pedicel region has two lateral humps marked by egg-shaped white spots. The legs are annulate. The basal two-thirds of the femur of all legs is orange in color while the distal one-third is black. The remainder of the legs have alternating black and white bands. The sternum is dark brown with a lighter, branched mid-longitudinal band. A taxonomic treatise of this species was given by Levi (1971).

FIGURE 72. A female *A. bicentenarius* displays its distinctive green abdomen with brown markings.

Ecology and Behavior: *A. bicentenarius* is found on trees in wooded areas. It is often found near greenish-colored lichens where it is superbly camouflaged. According to Levi (1971), a botanist found a specimen among lichens on jack pine. The greenish colored abdomen and the folium pattern tended to make the spider disappear among the lichen background. Another report indicated that a "green" specimen of *A. bicentenarius* was collected from a maple tree in West Virginia and that it blended perfectly with lichens. Most collections of this strikingly beautiful and rare spider consist of single specimens, very few of which are males. It is distributed in eastern North America from Nova Scotia, Minnesota to northern Florida and Texas (Levi, 1971).

Size: Length of female 13 to 28 mm; of male 7 mm (Levi, 1971).

Family Araneidae—*Araneus cavaticus* (Keyserling)—Barn Spider

Identifying Characteristics: The barn spider is a large, heavy-bodied species with well developed shoulder humps. It belong to a group of araneids known as the "angulate"orb-weavers, distinguishing them from the more "round-shouldered" forms that lack the shoulder humps. The barn spiders are different from most araneids in that the adults lack a well-defined folium on the abdomen. Rather, the overall pattern varies from pale bluish gray to greenish gray to grayish brown. Adult males and females are densely covered with spines and hair-like bristles.

Ecology and Behavior: This spider is more common in the northern than in the southern states. It makes its webs around human structures such as barns, bridges, arbors, fences and porches. It has also been found beneath overhanging cliffs (Kaston, 1948). In northern Vermont, we found that during the day the female remains in a loose silken retreat at the periphery of her web. At night, she spends time in her nearby web, often sitting near or at the web's hub. The large, long-legged male is usually nearby at the periphery of the web. This species matures and mates during July and usually attaches it egg sac above its web (Kaston, 1948).

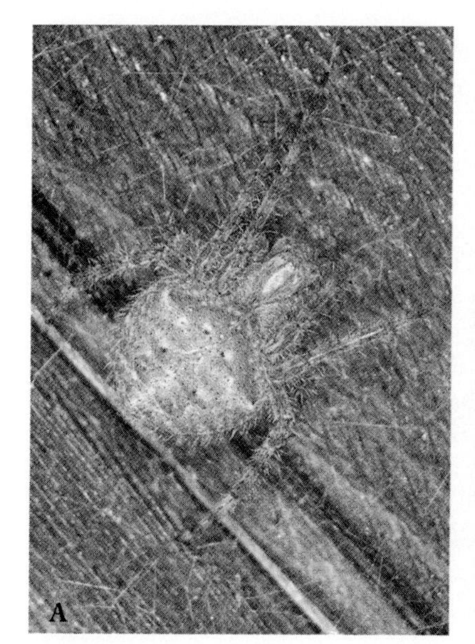
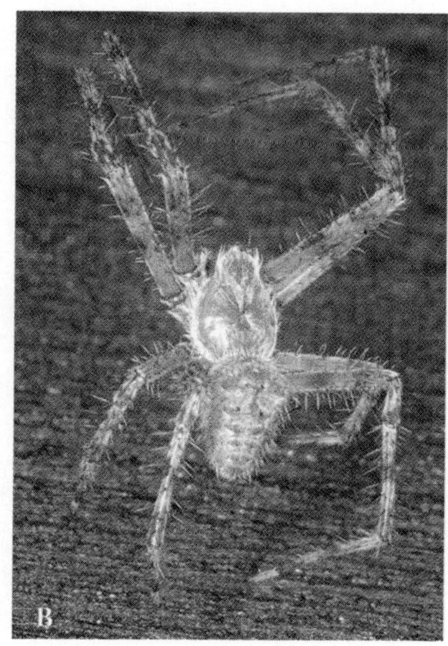

FIGURE 74 A-B. Female (**A**) and male (**B**) barn spiders.

This species ranges from the northeastern U.S. and adjacent Canada southwest through West Virginia to Alabama and Texas (Kaston, 1978).

Size: Length of female 13 to 22 mm; of male 10 to 19 mm (Kaston, 1978).

Family Araneidae—*Araneus cingulatus* (Walckenaer)

Identifying Characteristics: This small orbweaver has a rather homogeneous yellow-green background color over the carapace, sternum, and legs. The legs darken a little to a brownish color toward the ends. The ground color of the abdomen is also greenish-yellow. However, it has a pair of yellow spots anteriorly and two lateral yellow bands running the

FIGURE 75. A female *A. cingulatus* looking for a spot to anchor her web.

length of the abdomen. The lateral yellow bands each have about four to five embedded red dots equally spaced along the length of the abdomen. *A. cingulatus* is distinguished from the similar *A. niveus* by lacking the abdominal black patch (Kaston, 1978). This species is distinguished from other closely related species by the anatomy of the genitalia.

Ecology and Behavior: Most northern collections of this spider come from wasp nests. From Florida most have come from orange trees. The photographed specimen was collected from a deciduous forest in Alabama. Both sexes have been found during warmer months; May to July in northern states and March to September in southern states (Levi, 1973). *A. cingulatus* ranges throughout the eastern states, extending from Florida north to Massachusetts and west to Missouri and Texas (Kaston, 1978).

Size: Length of female 4.6 to 6 mm; of male 2.7 to 3.5 mm (Kaston, 1978).

Family Araneidae—*Araneus detrimentosus* (O.P.-Cambridge)

Identifying Characteristics: The bright green to chartreuse colored abdomen is perhaps the most prominent feature of this small orb-weaver spider. The green abdominal mark is narrow anteriorly near the pedicel but broadens posteriorly to cover most of the abdominal dorsum. Laterally, the green is bordered with a white stripe. Further laterally, the abdomen is brown. Ventrally, the abdomen is light brown with four white, indistinct patches. The abdomen is wider than long. The legs and carapace are brown. The carapace is covered with white pubescence. Jackman (1997) reported that the body coloration of the male is similar to that of the female.

Ecology and Behavior: *A. detrimentosus* spins a small orb web 80 to 150 mm in diameter with a silken canopy above. The spider is found in tall grasses, reeds, defoliated bushes, and on pine, juniper and elm trees.

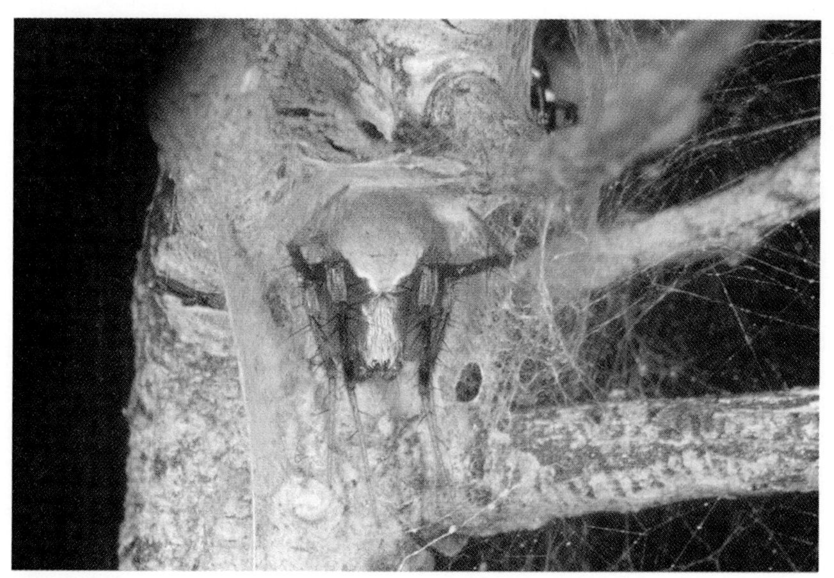

FIGURE 73. Female *Araneus detrimentosus* guarding her egg sac.

During the day, the spider hides itself in a silken retreat. This orb weaver is a southern, mostly coastal species that extends from Florida to California. Adults first appear in March and are found through October.

Size: Length of female up to 5.3 mm; of male 3.8 mm (Levi, 1973; Jackman, 1997).

Family Araneidae—*Araneus diadematus* (Clerck)—Cross Spider (Garden Spider of Europe)

Identifying Characteristics: This rather large orbweaver is also referred to as *Aranea trifolium* by Comstock (1912), but is definitely distinctive from that species. This spider may take on a variation in basal coloration, darkening with age. The general body color varies from a pale yellow to a dark brown. On the abdomen the folium is a darker shade of brown

FIGURE 76. A female Cross Spider, *Araneus diadematus.*

being more prominent among the younger females. Within the darker folium is an arrangement of white or yellow spots in the shape of a cross, attributing to the common name of "Cross Spider". This spotted cross is more obvious among the darker varieties and older specimens. On the carapace there is a median light border flanked with darker bands. The legs are banded in light and dark colors.

Ecology and Behavior: *Araneus diadematus* builds large and complete orb webs in the open sections of wooded areas or tall grasses. Generally, in daylight hours, the spider is found awaiting its prey from a retreat within a folded leaf at the periphery of the web, yet with a trap line intact with the hub of the web (Comstock, 1912). Alternately, the spider may be found in the hub of its web, especially at night. The Cross Spider ranges throughout the eastern states and west to Washington and Oregon. It is more common in northern states than in southern ones.

Size: Length of female 6 to 20 mm; of male 5.5 to 13 mm (Kaston, 1978).

Family Araneidae—*Araneus guttulatus* (Walckenaer)

Identifying Characteristics: *Araneus guttulatus* is a small spider with a yellow-green cephalothorax. The abdomen is distictively colored with a bright green, red and white transverse bands. Four pairs of red dots extend from the anterior shoulders to the posterior margins of the abdomen. The legs share the dull green of the cephalothorax. The distal ends of the femur and tibia of legs I and II are terminated with dull red bands. The color description given here for the photographed specimen may vary somewhat for other specimens of this same species. Pigmentation varies considerably from specimen to specimen in this tiny colorful spider. *A. guttulatus* is similar to another species *A. niveus*, but has more red on the abdominal dorsum and the base color is more greenish than that species (Kaston, 1978).

FIGURE 77. A male of *Araneus guttulatus*.

Ecology and Behavior: *A. guttulatus* occurs in a wide range of habitats. The photographed specimen was collected from a shrub along a southern marine salt marsh. The species also frequents open temperate forests and grasslands throughout the eastern U.S. It extends from New England south to Georgia and west to Arkansas and Wisconsin (Kaston, 1978).

Size: Length of female 3.8 to 6 mm; of male 3.9 to 4.8 mm (Kaston, 1978).

Family Araneidae—*Araneus marmoreus* (Clerck)—Marbled Orbweaver

Identifying Characteristics: This is one of the most colorful of the orb-weaving spiders. The carapace is homogeneous light amber to orange in color. The same coloration extends down onto the legs as far as the patellae. The oval abdomen has a ground color of bright yellow above. Zig-zagging the length of the abdomen is a dark purplish-black central band with scalloped edges which encloses yellow spots. The sides are marbled with black and yellow spots and blotches. The midventral area of the abdomen is dark, being enclosed by a pair of bracket-shaped light marks. The legs of the female have amber to orange coxae, trochanters, femora and patellae. The remaining leg segments are white proximally with dark rings distally. The palpus is orange basally with alternating white and dark annuli distally. In males, the leg segments are all orange at the base and dark distally. Males are only one-half the size of adult females.

FIGURE 78. A beautifully colored female marbled orbweaver, *Araneus marmoreus.*

Ecology and Behavior: This handsome spider builds its vertical orb web chiefly in deciduous woodlands or at the edges of woodlands. The adults are usually found only from August through October. Most webs are found from three to ten feet above the ground, being built between the branches of trees and shrubs. Where the upper corner of the web is attached to a limb, the spider constructs a cone-shaped retreat made from leaves woven together with silk threads. During the daytime the spider usually remains hidden in its retreat, huddled up with its legs drawn in. In its retreat, the spider holds onto the free end of a signal line that runs to the hub of the web. According to Fitch (1963), its prey consists of larger flying insects such as cicadas, katydids and beetles. Egg sacs are usually produced during October. This is a widespread spider being found from Alaska, throughout Canada, the eastern U.S. west to Texas and North Dakota, then across the Northern Rocky Mountains to Washington and Oregon (Kaston, 1978).

FIGURE 79. An old female Marbled Orbweaver collected during November in north Alabama. Note that the abdomen coloration has turned orange. Younger specimens are bright yellow on the abdomen (see **Figure 78**).

Size: Length of female 9 to 18 mm; of male 5.9 to 9 mm (Kaston, 1978).

Family Araneidae—*Araneus miniatus* (Walckenaer)

Identifying Characteristics: *Araneus miniatus* is a small yellowish to brownish spider faintly marbled with red or brown coloration. Between the anterior shoulders of the abdomen is a transverse white or yellow band outlined in red. Along the midline of the abdomen is a transverse white to yellow irregular series of spots. Anterior to these transverse spots is a pair of smaller white or yellow spots. Posterior to the transverse band are two longitudinal rows of distinctive black spots extending down the abdominal dorsum toward the spinnerets. There are three to four pairs of these black spots, decreasing in size as they approach the spinnerets. The legs are light green with red bands at the joints.

FIGURE 80. A female *Araneus miniatus* guarding her egg sacs hidden in a previously folded palmetto leaf.

Ecology and Behavior: *Araneus miniatus* is a nocturnal orb weaver which constructs its web in shrubs in open woodlands. During the day, the spider hides nearby its web within a folded leaf secured with silken strands of webbing. Additional silken signal lines are attached from the folded leaf to the web. During June in warmer climates, the female deposits and guards two or three egg sacs in folded leaf retreats enveloped in strands of silk. This orb weaver may be found in most temperate forests along the eastern coastal states from New England to Florida. In the southern U.S., its range extends westward to Texas (Kaston, 1978).

Size: Length of female 3 to 7 mm; of male 2 to 3.7 mm (Kaston, 1978).

Family Araneidae—*Araneus pegnia* (Walckenaer)

Identifying Characteristics: This species has also been known as *Neosconella pegnia.* This tiny orbweaver has a small body relative to the leg length

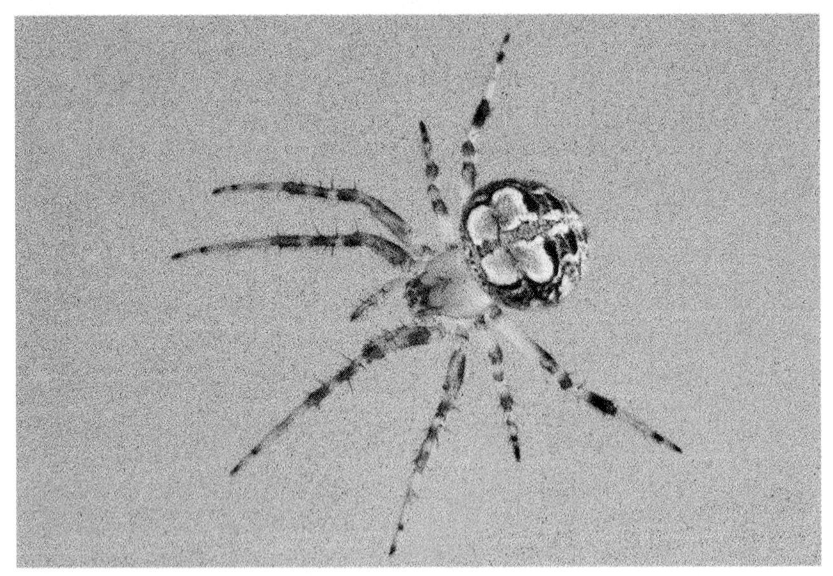

FIGURE 81. Female *Araneus pegnia* exhibiting the distinctive "butterfly-like" color pattern on the abdomen.

(tibia I plus patella I is 1.5 times the length of the carapace), an oval abdomen which rides above the cephalothorax, and a distinctive folial pattern on the abdomen. While the abdomen may have varying shades of brown, it does bear dorsal white markings forming an outline somewhat reminiscent of a swallowtail butterfly. Two pairs of anterior brown spots bordered in white resemble the wings of the butterfly, and a double white mid-dorsal line extending to the spider's spinnerets resembles the tail. The cephalothorax is light brown and possesses features that have been used to characterize the genus. The head region is not clearly set off from the thoracic portion of the cephalothorax. Most importantly, in the thoracic region there is a circular depression. The same head thoracic region is transversely depressed in *Araneus* and longitudinally depressed in *Neoscona*.

Ecology and Behavior: The snare of this spider consists of an orb web which has a free section lacking spiral strands and a leader used by the spider to sense vibrations from entrapped prey. Also, an irregular labyrinth of silk threads may be associated with the web, similar to those found in the web of *Metepeira labyrinthea*. Often, a small silken tent is woven into this mesh and the spider uses this as a retreat. This spider is common in the eastern states but may be found throughout the U.S. (Kaston, 1978).

Size: Length of female 3.5 to 8.2 mm; of male 2.5 to 5 mm (Kaston, 1978).

Family Araneidae—*Araneus trifolium* (Hentz)—Shamrock Spider

Identifying Characteristics: On the carapace, fine hairs cover and nearly obscure the flesh-colored head which has a central black stripe and another running along the margin of each side. The legs have alternating flesh-colored and black rings. According to Kaston (1978), the abdomen varies greatly in coloration from "pale green, or brown to gray, or even purplish red, but the pattern of light spots is always the same." The

FIGURE 82. A handsome shamrock spider, *Araneus trifolium,* pauses head downward on a blue flower after leaving its folded leaf retreat.

abdomen in our specimen from northern Vermont (see above) is pale green with white spots. The light spots on the abdominal dorsum nearly always conform to the pattern seen in the above figure, i.e., one anterio-medial spot flanked laterally by three or four smaller spots, followed by three coalescent medial spots, then two small and two large spots, then one medial spot, four tiny spots, a single medial spot, and another single medial spot.

Ecology and Behavior: These large orb-weaver spiders are found throughout the United States, but are more common in the cooler, more northernly states. The orb-web snares are usually placed in open areas among tall grasses. During the day time, the spider may be found at the edge of the web in a retreat formed by two or three folded leaves. The spider monitors vibrations within its web by means of a signal thread which is attached to the hub of the web. It prefers tall grasses, weeds and bushes in rather moist situations. Maturation takes place during July and August. Mating usually occurs in August. Eggs sacs are deposited during September and October. They are whitish, about an inch in diameter, and contain several hundred eggs (Kaston, 1948). The eggs usually hatch and young spiderlings emerge during the succeeding spring.

Size: Length of female 9 to 20 mm; of male 4.5 to 8 mm (Kaston, 1978).

Family Araneidae—*Araniella displicata* (Hentz)—Sixspotted Orbweaver

Identifying Characteristics: This spider has a very simplistic color pattern which distinguishes it almost immediately. The carapace and legs are yellowish-brown to orangish-brown without markings. The abdomen coloration may be yellow, white or pink. On the dorsal and posterior half of the abdomen are three pairs of small black spots. The median ocular

FIGURE 83. Female *Araniella displicata* showing its white abdomen with three pairs of black dots.

area is as wide behind as in front and the posterior median eyes are a little larger than the anterior median eyes. The palps of the male bear three spines on the tibia.

Ecology and Behavior: We have found adults of *Araniella displicata* during the month of May. The spider makes its very tiny web about 4 to 6 feet above ground among the lowest limbs and within rolled up leaves of sycamore trees. Often the spider is found hiding within the rolled up leaf. During the month of May, the females lay their egg cases within the rolled up leaves. The egg masses are covered with a fluffy, entangled mass of silken threads. A single egg sac contains about 85 eggs. Kaston (1948) observed this species in Connecticut and stated: "It builds a very small web, in tall grass and bushes; often in the space enclosed by the bending of a single leaf. No retreat is made, the spider standing at the hub of its snare. Individuals overwinter in the penultimate and younger instars mature in May. Mating occurs in late May and early June and egg sacs have been found in early July. One sac found July 4 was a fluffy transparent mass about 8 mm in diameter, attached to the underside of a leaf. It contained 78 greenish-brown non-agglutinate eggs, each measuring about 0.45 by 0.38 mm. It is a very widespread species occurring throughout most of the United States and Canada to Alaska."

Size: Length of female 4 to 8 mm; of male 4 to 6 mm (Kaston, 1978).

Family Araneidae—*Argiope argentata* (Fabricius)—Silvered Garden Spider (or Silver Argiope)

Identifying Characteristics: Different from other spiders of this genus, *A. argentata* is distinguished by its wider abdomen, nearly as wide as long. The anterior half of the abdomen is silvery white to yellow. Posteriorly, the abdomen is dark and bears a median lobe and three pairs of lateral lobes. The carapace is covered with silvery white hairs.

FIGURE 84. *Argiope argentata* sits in its web awaiting its next meal.

Ecology and Behavior: *A. argentata* is recognized in the field by the loose orb web generally constructed among tall grasses and shrubs in open areas. The spider rests in the center of the web, head up and with the first and second pairs of legs directed anteriorly and the third and fourth pair of legs directed posteriorly. In line with each pair of legs is a stabilimentum, which appears as a cross with the spider's legs forming the center of the cross. *A. argentata* is a tropical and semitropical species, common from Florida westward across the southern states to California.

Size: Length of female 12 to 16 mm; of male 3.7 to 4.7 mm (Kaston, 1978).

Family Araneidae—*Argiope aurantia* (Lucas)—Black and Yellow Garden Spider

Identifying Characteristics: This is one of the largest, most conspicuous, and most familiar of eastern orb weaving spiders. The female has the carapace covered with silvery-white hairs. The abdomen is somewhat oval in shape, pointed behind and notched anteriorly to form two small dorsolateral, tubercle-like "shoulder-humps." The abdomen is black and yellow, sometimes yellowish-orange. The abdomen bears a middorsal black stripe which is narrowest between the shoulder humps and at the posterior end. It is widest in the middle portion of the abdomen and encloses two pairs of yellow spots, the anterior spots being the larger. Along the sides are about five pairs of bright yellow bands. The venter is black with a yellow stripe on the sternum and two wide yellow stripes on the abdomen. Small yellow spots are located between and along the

FIGURE 85. A female *Argiope aurantia* in her web.

sides of these yellow stripes. The legs usually have alternating black and yellow-orange annuli.

Young females are distinctly different from the adults in color pattern by having legs with dark rings on the ends and middle of each joint. Additionally, the abdomen is narrower and more slender, the color is pale, the markings are gray, and the strong black and yellow of the adults are lacking. The adult male is similarly colored as the female but is only about one-fourth the length. The markings are less distinct and the palpi are very large.

Ecology and Behavior: The young spiders first become evident during late June or July when they construct a small orb web less than 30 cm in diameter and with an inordinately wide and short zig-zag stabilimentum (up to 6 cm in width) in the center of the orb. The adult female makes a large orb web, up to 60 cm in diameter with a long and narrow stabilimentum (up to 3 cm in width and 35 cm in length). The web is slightly

FIGURE 86. Immature female *Argiope aurantia* in her web. Note the exceptionally wide stabilimentum characteristically built by immature black and yellow garden spiders.

inclined, and the spider usually remains in the center of the web on the lower side with its head pointed downward. The males make small imperfect webs of their own near the webs of the females. In September, the females lay their eggs in large (up to 2.5 cm in length), tough, brownish, paper-like and pear-shaped cocoons, which are hung by a meshwork of tough threads among grasses and limbs of small shrubs and trees. The adults die during October around the time of the first frost. The young spiders hatch during the winter and emerge from the cocoon during May. This spider is perhaps one of the best known spiders in the south because of its abundance, bright colors, large size and habit of building its webs in gardens and flower beds around human habitations.

Size: Length of adult females 19 to 28 mm; of males 5 to 8 mm (Kaston, 1978).

Family Araneidae—*Argiope trifasciata* (Forskal)—Banded Garden Spider

Identifying Characteristics: In the female, the dorsal ground color is predominantly white to pale yellow. Silvery-gray hairs are present over the carapace. A pair of dark blotches is found on the sides of the thorax. On the dorsum of the abdomen, 12 to 15 thin, silver and yellow transverse bands alternate with black bands. Some of these bands are complete, while others are broken. The legs are amber in color with black rings. Legs I have black femora. The palps are amber in color. The ground color of the venter is black. The sternum has a bright yellow longitudinal area along its midventral portion. Yellow parallel stripes are present on the abdomen, each broken near its posterior end. There are three pairs of white dots set against the black mid-ventral region of the abdomen. A series of transverse grooves is apparent on the posterior portion of the abdominal dorsum. These grooves correspond in position to the transverse black bands and give the abdomen a segmented appearance. The posterior end of the abdomen is more pointed than that of its relative, *Argiope aurantia.* According to Fitch (1963) the males are only one-

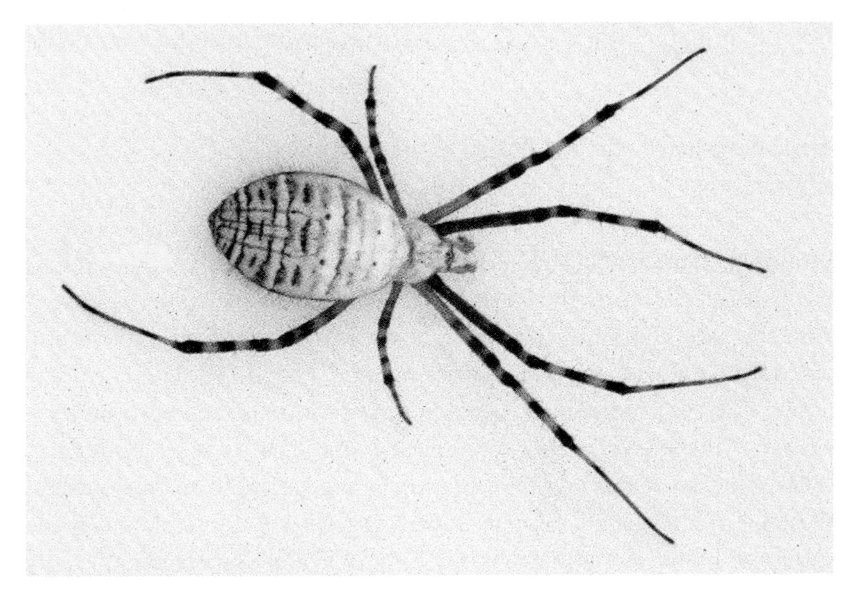

FIGURE 87. Female *Argiope trifasciata.*

fourth the size of the females, the abdomen is much less bulky, and almost entirely white over the dorsum.

Ecology and Behavior: This common, large orbweaver builds its web in grasses and low bushes in sunny areas. The stabilimentum is similar to that of *Argiope aurantia.* Throughout the spring and part of the summer these spiders are still young and inconspicuous. In late August, many specimens become sexually mature. During late August and September, a male and a female can be seen together in the same web. The females will drop from the web to the ground at the slightest disturbance, whereas the male will usually remain in the web. From September to October a large hemispherical, or kettledrum-shaped cocoon is fastened to the top of a weed or grass stem near the web. Most of the adults die when the first frost comes. This well-known spider is found throughout most of the U.S. and in warmer regions worldwide.

Size: Length of female 15 to 25 mm; of male 4 to 5.5 mm (Kaston, 1978).

Family Araneidae—*Cyclosa conica* (Pallas)—Conical Trashline Orbweaver

Identifying Characteristics: This small spider is distinguished by a conical hump at the posterior end of the abdomen. When the spider is viewed from the side, the hump is so pronounced that it extends upward and backward and well beyond the spinnerets. The hump varies in size among different individuals, becoming more pronounced with increasing age. Males have a less pronounced hump than females (Emerton, 1961). The amount of color variation is considerable in this species. Most specimens have a mixture of drab gray and white markings, while some specimens are almost black. Lighter colored specimens usually have some black markings along the midline of the abdominal dorsum.

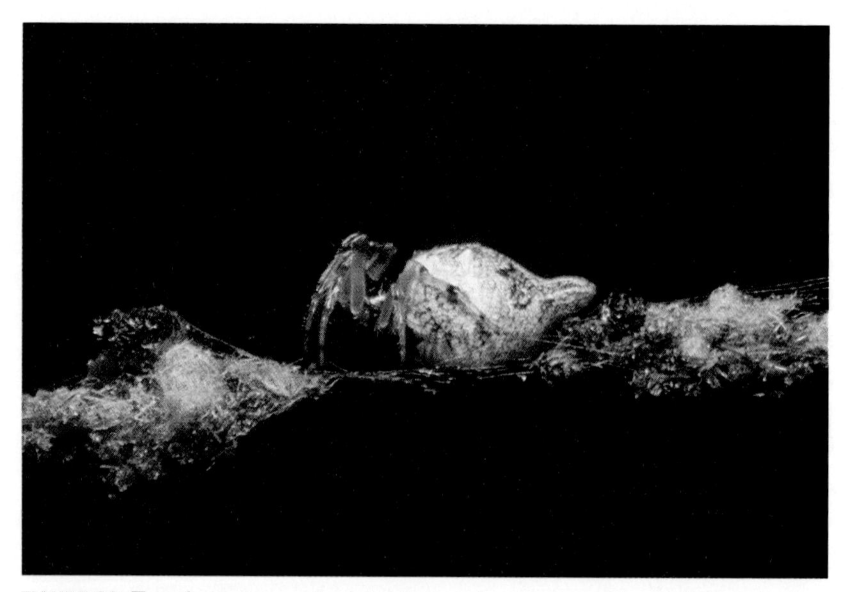

FIGURE 88. Female *Cyclosa conica,* lateral view, showing posterior conical projection.

The venter is black except for a pair of very distinct light spots across the middle. The cephalothorax is dark gray to black. Except for the femurs, the legs are light, with dark annular rings at the end of each joint and in the middle of each leg segment.

Ecology and Behavior: This spider makes a very small vertical orb web in open woodlands and in ornamental shrubs around human dwellings. The spider lives in the hub of the web and rarely leaves the web. The remains of prey and the swathed prey are usually deposited along the stabilimentum both above and below the hub. The deposits are about the same size and color of the spider, effectively camouflaging the spider's presence. The adult males do not build webs. The sexes mature in spring, and after mating, the female constructs up to five egg sacs and deposits them on dead twigs or under leaves.

Size: Length of female 5.3 to 7.5 mm; of male 3.6 to 4 mm (Kaston, 1978).

FIGURE 89. The web of *Cyclosa conica* showing the deposits of prey along the stabilimentum. The tiny camouflaged spider sits head downward at the center of the web.

Family Araneidae—*Eustala anastera* (Walckenaer)

Identifying Characteristics: The carapace is usually grayish in the midline but becomes darker laterally. The abdomen is somewhat triangular in shape, being widest anteriorly and tapering posteriorly where it becomes bluntly pointed. The color patterns on the dorsum of the abdomen distinguish this species, but they also display a great deal of distinct variations. Comstock (1912) diagrammed four rather distinctive variants in abdominal coloration. Although there is much variation in abdominal color patterns, the most common coloration is mottled gray to brown along the sides with a darker, wide, scalloped edged triangle coursing down the middorsum. This colored triangle is widest anteriorly, tapering to a point at the end of the abdomen. The legs are marked with spots and rings. The scape of the epigynum is a unique tapered structure with transverse wrinkles (Comstock, 1912).

FIGURE 90. A female *Eustala anastera* camouflaged against a lichen-colored tree trunk.

Ecology and Behavior: Comstock (1912) felt that the color patterns were very protective of this species. The color patterns often resemble the bark of the plant on which the spider rests. When disturbed the spider quickly runs to that part of the bark that best camouflages it. It then pulls in its legs and crouches down to blend in with the mottled or lichen-covered bark. The orb web is vertical and constructed in low trees and bushes. This common orbweaver is found from New England south to Florida and westward to the Rocky Mountains (Kaston, 1978).

Size: Length of female 5.4 to 8 mm; of male 4 to 6 mm (Kaston, 1978).

Family Araneidae—*Gasteracantha cancriformis* (Linnaeus)—Spinybacked Orbweaver

Identifying Characteristics: This spider is also known as *Gasteracantha elipsoides* (Walckenaer). This orb weaver is recognized by the bright

FIGURE 91. Female *Gasteracantha cancriformis* showing its unique spiny abdomen.

yellow, white or orange abdomen with a hardened enameled appearance and several evenly distributed oval spots on its dorsal surface. The raised dorsum of the abdomen is held transversely, being much wider than long, with four prominent spurs at each corner. Two spurs project from the abdomen posteriorly. The spurs are bright red or black and are sharply pointed. The cephalothorax is dark brown and short, being as wide as long. The entire venter is black with small yellow spots. Common to its subfamily, Gasteracanthinae, the spinnerets are elevated on a large projection and surrounded by a thick ring.

Ecology and Behavior: The orb webs, built between three and six feet from the ground in open wooded areas, are supported by foundation lines anchored to the lower branches of trees and bushes. The web of the female can reach a diameter of 40 cm, and it usually contains the adult resting in the hub. *G. cancriformis* is restricted in its range to the southern states. It extends from North Carolina south to Florida and west to California.

Size: Length of female 8 to 10 mm; of male 2 to 3 mm (Kaston, 1978).

Family Araneidae—*Gea heptagon* (Hentz)

Identifying Characteristics: The carapace is pale yellow with dark brown markings between the radial furrows. The dorsum of the abdomen is yellow with a darker triangular shaped marking near the posterior end, as well as other scattered dark marks. The legs are yellowish with brown annuli. Tibia I of the male is curved with many bristles. A pair of prominent dorsolateral tubercles are seen on the anterior portion of the abdomen.

Ecology and Behavior: According to Kaston (1978), this small spider sits in the center of its orb web which occasionally has a sector missing from its lower portion. The snare is usually constructed in vegetation near the ground, and it lacks a stabilimentum. According to Nyffeler et al.

FIGURE 92. A female *Gea heptagon* suspended in the conventional "head-down position" in her web.

(1989), aphids made up nearly half of all insects trapped in the snares of this species. When this spider is frightened, it drops from its web and changes color by becoming darker in the lighter regions of the body. This rapid color change helps it to blend in with the ground and become more difficult for a predator to see. The egg sacs are ivory in color, flattened, and usually contain from 30 to 45 eggs (Sabath, 1969). This species is found from New Jersey to Michigan and south to Florida and Texas and on to California (Kaston, 1978).

Size: Length of female 4.5 to 5.8 mm; of male 2.6 to 4.3 mm (Kaston, 1978).

Family Araneidae—*Larinia directa* (Hentz)

Identifying Characteristics: This unique spider has its abdomen more than twice as long as wide. The anterior portion of the abdomen forms a cone-like structure that projects slightly over the carapace. While at rest, the spider sits with legs I and II directed straight forward, while legs III and IV are directed straight backward. It can sit well-camouflaged on the stem of a plant or the blade of a grass. It is particularly well-camouflaged when it inhabits broom sage grasses. Its carapace is yellow with a median brown line extending from the median eyes posteriorly to the pedicel. The abdomen is yellowish-orange with several longitudinal stripes extending the length of the dorsum. Two light yellow stripes course down the middorsum and enclose a narrow, pinkish stripe. Lateral, and to each side of this middorsal stripe are parallel, longitudinal stripes that are, in sequence, brown, yellow, pinkish, yellow, and brown in color. Two rows of six black spots extend down the abdominal dorsum. All four pairs of legs are pale yellow in color with the exception of discrete black spots that are widely and equally spaced.

Ecology and Behavior: This spider is often found in sunny, grassy localities. It constructs an obliquely oriented orb web. As indicated above, it often rests on stems of plants and blades of grasses where it is well-camouflaged. Its habit of stretching legs I and II straight forward and legs III and IV straight backward when resting on plants is reminiscent of the habits of the tetragnath spiders. It is so well-hidden that it is most often overlooked by spider collectors. It is most frequently captured by using a sweeping net through fields of medium to tall grasses. It is distributed throughout most of the eastern U.S. and west through the southern states to California (Kaston, 1978).

Size: Length of female 4.8 to 11.7 mm; of male 4.5 to 6.5 mm (Kaston, 1978).

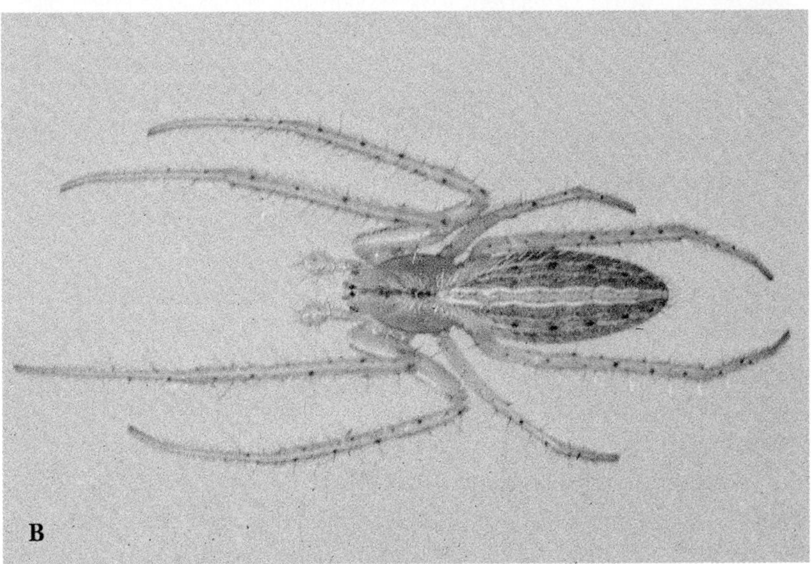

FIGURE 93 A-B. (**A**) female *L. directa* camouflaged in a leaf; (**B**) male *L. directa*.

FIGURE 94. This web of *Larinia directa* was supported by the leaves of a yucca plant. The adult spider rested nearby in a folded leaf of the same plant.

Family Araneidae—*Larinioides cornutus* (Clerck)—Furrow or Foliate Orbweaver

Identifying Characteristics: This common spider has long been known by the name *Nuctenea cornuta,* but the name was recently changed (Levi, 1974). The common name, foliate orbweaver, refers to the distinctive pattern on the dorsal abdomen. In the female the oval elongate abdomen is light grayish brown with a wide medial darker folium. The wavy margins of the folium are entire, rarely broken. Along the midline of the abdomen are lighter markings and anteriorly is a small black triangular cardiac mark. Ventrally, the abdomen is black with a pair of white comma-shaped marks. The cephalothorax is grayish to reddish brown with a dark stripe on the lateral margins. The cephalothorax is heavily covered with bristly hairs. The slightly smaller male resembles the female.

FIGURE 95. Female *Larinioides cornutus* exhibiting its distinctive pigmentation on the dorsum of the abdomen.

Ecology and Behavior: In the eastern states, *L. cornutus* is the most common araneid house spider. There are three species of *Larinioides* throughout North America and all three are more commonly found about houses, barns, and fences than are other orb-weavers. The snare is a moderately tight orb that often exceeds three feet in diameter and is supported by vegetation and man-made structures. During daylight hours these spiders frequently are hidden in a silken retreat at the periphery of the web. However, during a calm evening the female will often rest in the web awaiting her prey. The photographed specimen was collected from her web in the railings of a well traveled foot bridge where the spiders seemed unperturbed by the heavy attention from evening walkers fascinated by their active web weaving abilities. *L. cornutus* is found throughout North American but it is most common east of the Mississippi River. The other species of *Larinioides* are most common in northern and western states (Levi, 1978).

Size: Length of females is 6.5 to 14 mm; of males is 4.7 to 9 mm (Kaston, 1978).

Family Araneidae—*Larinioides patagiatus* (Clerck)

Identifying Characteristics: *L. patagiatus* is similar to *L. cornutus* in general appearance and is often confused with that species. The two forms may be distinguished for certain only by microscopic examination of the female genitalia.. The epigynum is furrowed posteriorly at its base and its scape has a narrow neck. However, some color pattern differences do seem to separate the two species and one does not have to resort to such rigorous microscopic analysis to distinguish the species in the field. As in *L. cornutus,* the abdomen is somewhat dorsoventally flattened, widest about the middle of its length. The midventer of the abdomen is black with a white, comma-shaped spot on each side. Dorsally, the folium is very distinctive being dark brown to black with a dark cardiac mark

FIGURE 96. Female of *Larinioides patagiatus.*

(Kaston, 1978). The folium has a whitish, mid-dorsal stripe that extends from the cardiac mark posteriorly about three-fourths the length of the folium. This light stripe is crossed anteriorly by two white bands. These bands are normally wider and broader that those in *L. cornutus.* The carapace is brown and covered with many fine hairs.

Ecology and Behavior: *L. patagiatus* is a nocturnal orb weaver that commonly constructs webs around manmade structures, especially barns, houses and bridges. It is most common around permanent bodies of water, such as lakes and ponds. During the daylight hours, *L. patagiatus* often remains in its silken, cocoon-like retreat near the site where it will construct its web. It leaves it retreat and constructs its web at nightfall. Mature males and females may be found at all seasons of the year. This species is more common in the northern states. Its range extends from New England and adjacent Canada south to North Carolina and west to the Pacific northwest states. It also occurs in Arizona.

Size: Length of female 5.5 to 11 mm; of male 5 to 7 mm (Kaston, 1978).

Family Araneidae—*Larinioides sclopetarius* (Clerck)—Gray Cross Spider

Identifying Characteristics: The gray cross spider is the largest and darkest of the three members of *Larinioides* in North America. The border of the carapace in this species is uniquely trimmed in white hairs. The abdominal folium is outlined in white. On the dorsal and anteriomost portion of the adbomen is a white "V" shaped marking with the tip of the "V" pointed toward the head. In the region of the second abdominal segment the white folial boundary is broken and crossed with a lighter gray patch. The front legs of *L. sclopetarius* are longer than those of the other members of the genus.

Ecology and Behavior: *L. sclopetarius* is Eurasian in distribution and in North America is most abundant in the Great Lakes states. It is the least

FIGURE 97. Male of *Larinioides sclopetarius.*

frequently encountered species of the genus. Based upon this, Levi (1978) suspected that *L. sclopetarius* was probably introduced into North America. When encountered, the web is generally constructed around buildings and bridges near bodies of water. The spider does not construct a peripheral retreat but waits at the end of a radial strand. It ranges from Canada south to North Carolina and west to the Pacific coast.

Size: Length of female 8 to 14 mm; of male 5.5 to 8.5 mm (Kaston, 1978).

Family Araneidae—*Mangora gibberosa* (Hentz)—Lined Orbweaver

Identifying Characteristics: This spider is similar in size, shape and color to the other members of the genus *Mangora*. The carapace, legs, and abdomen share a common background color that is a pale yellow or, less common, a pale green. Ventrally, the abdomen is brown with white spots.

This spider is distinguished from the other spiders of the genus by three markings. First, along the midline of the carapace is a thin black line extending posteriorly from the thoracic groove to a point anteriorly between the eyes. Second, femora I and II have a thin longitudinal black line on their undersides. Third, the abdomen has several black spots anteriorly and a pair of longitudinal black bands which extend posteriorly to the spinnerets. On some individuals these two lines appear dotted.

Ecology and Behavior: Much like other members of its genus, it constructs a tightly woven web with a large number of radii (50 to 60) and viscid spirals. The web is built slightly inclined from the horizontal in low bushes and tall grasses. The adult usually rests in the hub or it seeks cover in a nearby curled leaf. Adults mature during July and August. Eggs from the female are deposited in cocoons within a small folded leaf. The eggs hatch in the fall; however, the spiderlings do not emerge from the cocoons until the following spring (Kaston, 1948). This spider is the most common member of its genus in the northern U.S., but is less

FIGURE 98. Female *Mangora gibberosa.*

common than *M. placida* in the southern states (Kaston, 1948). Its range extends from Canada to Florida and west to the Rocky Mountains

Size: Length of female 3.4 to 6 mm; of male 2.6 to 3.2 mm (Kaston, 1978).

Family Araneidae—*Mangora maculata* (Keyserling)—Greenlegged Orbweaver

Identifying Characteristics: *M. maculata* is similar in size, shape and color to the other species of its genus. Its carapace and legs are dull yellow. The black lines found on the carapace and on femora I and II in *M. gibberosa* are not present. The dorsum of the abdomen is a dull yellow to white except for three pairs of taxonomically distinguishing black dots arranged in two rows along its posterior end.

FIGURE 99. Male *Mangora maculata.*

Ecology and Behavior: Webs are tightly woven and constructed in an almost horizontal plane. They are found in low bushes and grasses in hardwood forests. Adults are generally found in the hub of the web. They reach sexual maturity in June and July, and deposit eggs in the fall in cocoons wrapped in dried leaves. *M. maculata* is much less common than are *M. gibberosa* and *M. placida*. Its greatest numbers occur from Texas to the southern states on the Atlantic seaboard. Its range extends into the New England states; however, it is much less common there.

Size: Length of female 3.6 to 5.5 mm; of male 2.7 to 4 mm (Kaston, 1978).

Family Araneidae—*Mangora placida* (Hentz)—Tuftlegged Orbweaver

Identifying Characteristics: The carapace is yellowish to greenish yellow with a brown stripe along each side and down the middorsum. The

FIGURE 100. Female *Mangora placida*.

dorsum of the abdomen is whitish except for the black sides and a conspicuous dark brown mid-dorsal stripe which widens posteriorly. The brown mid-dorsal stripe is broken posteriorly by a pair of white dots and white lines. The legs are green with faint, dark annuli. Tibia III bears a double series of long thin feathery hairs on its prolateral surface, a character typical of the genus. A good hand lens or a dissecting microscope is needed to see this character.

Ecology and Behavior: *M. placida* builds a tightly spun web about 30 cm in diameter and in the vertical or near vertical position. In the center of the web is an open hub in which the adult typically rests while awaiting its prey. Webs are commonly found in wooded areas with thick undergrowths of bushes, vines, and small trees. They can also be found in tall grasses. It even ventures into the suburbs where it will build its snare on shrubs of gardens and lawns. Archer (1940) found mature females in winter, spring and summer and stated that it was hard to assign any particular season for the reproduction of the species. He declared that this spider was the commonest orbweaver in cities in Alabama during the winter of 1939, but in the winter of 1940, it was very scarce. *M. placida* generally overwinters as young instars. The adults can be among the first spiders to appear in the spring. Specimens in the south have been collected as early as February and as late as October. This species is widely distributed, extending throughout most states east of the Rocky Mountains.

Size: Length of female 2.3 mm to 4.5 mm; of male 2 to 2.8 mm (Kaston, 1978)

Family Araneidae—*Mastophora bisaccata* (Emerton)

Identifying Characteristics: This small stumpy spider is easily recognized by its yellow color with markings of red or red-brown and two prominent horns extending from the posterior cephalothorax. The cephalo-

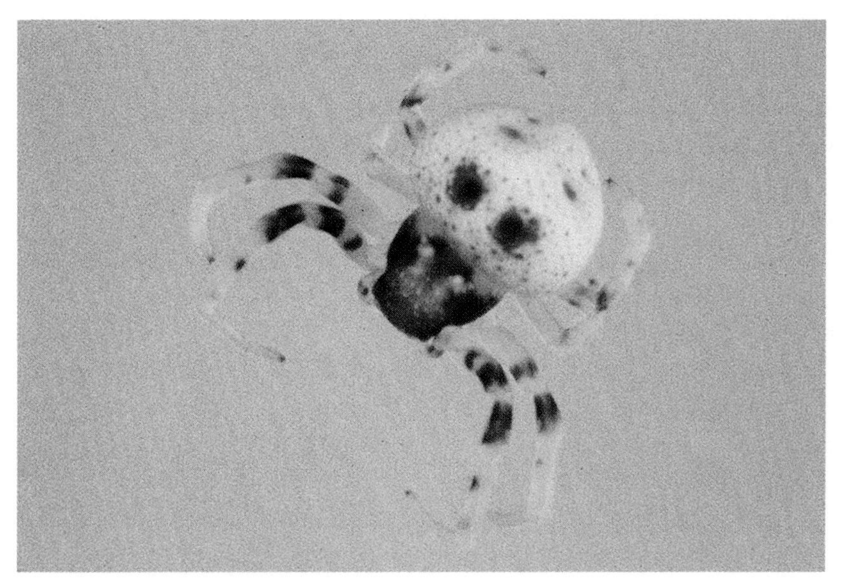

FIGURE 101. Female *Mastophora bisaccata*

thorax is slightly scalloped on the sides and is steeply sloped dorsally from the ocular quadrangle to the horns. The cephalothorax is more prominently marked in red and brown anteriorly and is covered with numerous irregular bumps. The abdomen is wider than long and almost as wide as the length of the body. The abdomen is yellow scattered with red or brown specks and has four deep muscle impressions that are red or brown. The legs are yellow like the rest of the body except legs I and II, which are ringed in red or brown. The venter, sternum and abdomen are yellow.

Ecology and Behavior: *Mastophora bisaccata* is a kind of "bolas spider" which does not make a web. Instead, at dusk, it strings a horizontal silk line to the underside of a twig. The spider then hangs from this trapeze-like line, and drops from her spinnerets a single line of silk about 50 mm long (2 inches). A drop of sticky, gum-like substance is attached to the end of the silken line. This serves as the spider's "fishing line" or "bolas". The spider's bait is thought to be a drop of pheromone that imitates that of certain female moths. This attracts the male moths and they become entangled in the swinging silken line which is intentionally manipulated by the spider itself. The spider pulls the entangled moth closer and closer until it is near enough for her to bite the moth and paralyze it. The moth is wrapped in silk and eventually eaten. *Mastophora* produces an egg sac suspended by a silken thread from its substrate. Kaston (1920) discovered its egg sac in an apple tree and Emerton (1919) found the egg sacs from oak and beech trees. Females mature in late summer to early fall. This species was described as a northern species (Comstock, 1912) and also been reported from Texas (Jackman, 1997). Roth (1993) listed five species in this genus with their distribution in the eastern U.S. from Maine to Minnesota and Kansas, south to Florida and across the southern states to California. The photographed specimen was collected in north Alabama.

Size: Length of female 6.5 to 8 mm; of male 1.5 to 3 mm.

Family Araneidae—*Mastophora phrynosoma* (Gertsch)—Bolas Spider

Identifying Characteristics: This species of bolas spider has a carapace that is mostly beige to flesh colored except on the posterior midline where a lighter somewhat squarish design is seen from above. This lighter area is mostly occupied by two, bifurcated and elevated horn-like structures. When seen head-on, these horn-like structures are very prominent and the spider strangely resembles the toads of the South American genus *Phrynosoma*. The median eyes are dark brown. The abdomen is triangle shaped, being very wide anteriorly, nearly three times the width of the carapace, and tapering to a bluntly rounded point posteriorly. The anterior two-thirds of the abdomen is beige to flesh-colored like the carapace, while the posterior one-third is cream-colored to white. The cephalothorax has many tiny wart-like bumps in addition to the two horns. The anterior two-thirds of the abdomen is wrinkled, pitted and

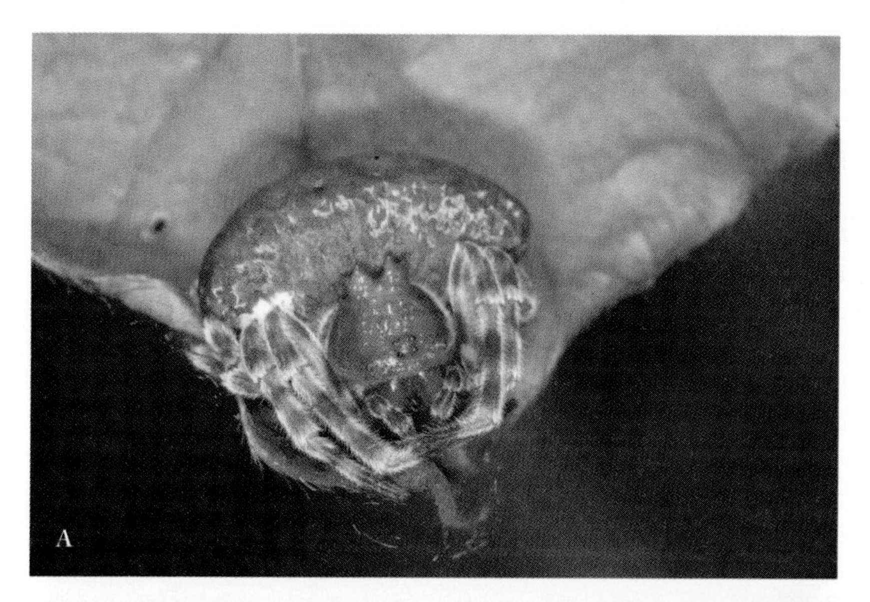

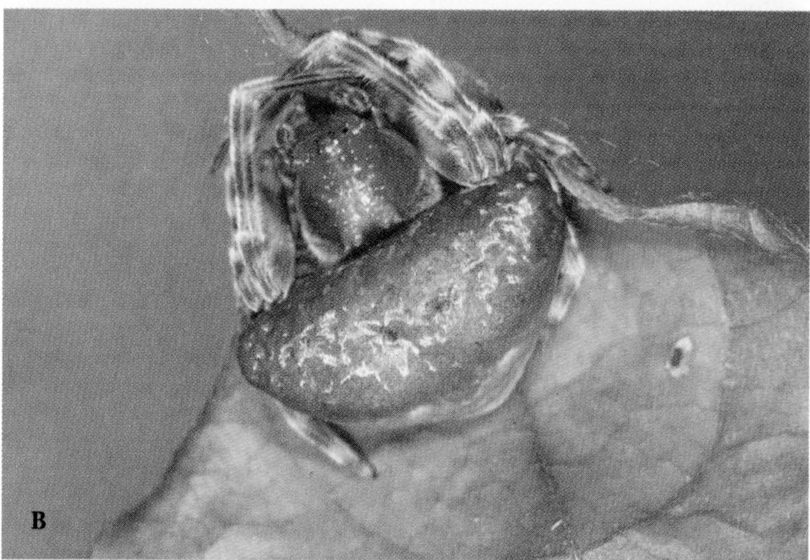

FIGURE 102 A-B. A bolas spider. (**A**) Head-on view; and, (**B**) dorsal view.

wart-like. The whole spider, except the legs, is shiny and appears wet as if it were a fresh bird dropping. All legs are ringed with dark gray annuli that are partially obscured by a dense, fine, covering of short, white hairs.

Ecology and Behavior: Bolas spiders are strange spiders in several ways. First, they are seldom seen by humans. This is because they are active only at night. Second, they are very odd in their shape, often resembling a bird dropping. And, finally, they get their name by hunting with sticky balls on the end on a silken line and resemble the South American gauchos throwing their bolas. They hang on a horizontal silk line and hold the bolas line with one leg. It swings the bolas at certain kinds of moths that are drawn to the bolas by a pheromone sex attractant. The moths are stuck to the bolas by a sticky chemical adhesive. Bolas spiders are found from New Hampshire to Minnesota to the southern states and west to California.

Size: Length of female 7 to 9 mm in length; of male 1 to 2 mm.

Family Araneidae—*Mecynogea lemniscata* (Walckenaer)—Basilica Orbweaver

Identifying Characteristics: *Mecynogea lemniscata* is easily recognized as the yellow-green and orange spider hanging upside down in a domed (basilica) web. Its cephalothorax is yellow to light brown with a dark narrow line running down its middle, and a wider dark band around its edges. The legs are an olive green color and possess numerous spines. Taxonomically important are the long tarsi and metatarsi which are longer than the tibiae and patellae. Also characteristic of this spider is the elongate abdomen, which is at least twice as long as wide and forms an anterior hump overhanging the carapace. Laterally, the abdomen is olive green. Dorsally, the abdomen has a broad white longitudinal band with jagged edges that may be bordered by red-orange or brown. Along

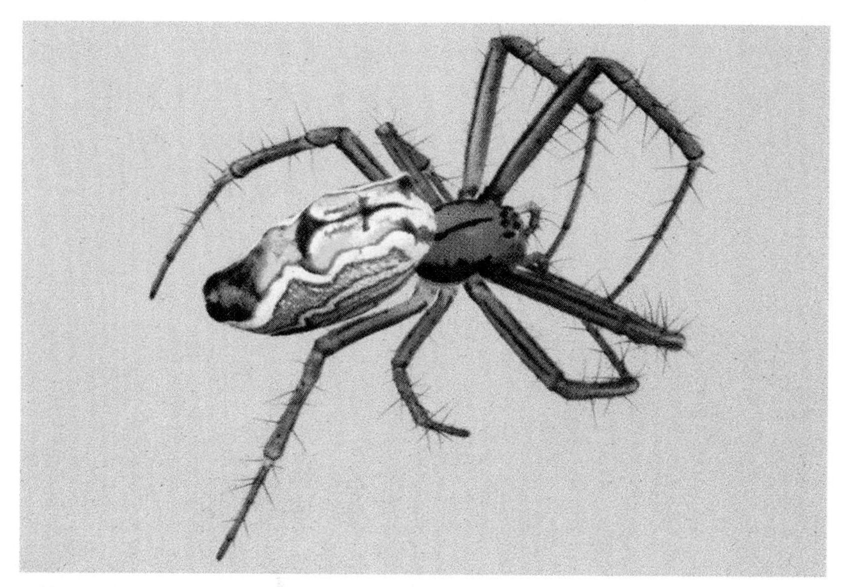

FIGURE 103. Female *Mecynogea lemniscata.*

the middorsum of the abdomen are variably shaped folial markings in black and yellow. The photographed specimen has a prominent cross and pyramid shaped markings on its dorsal abdomen.

Ecology and Behavior: *Mecynogea lemniscata* has acquired the common name, "basilica orbweaver" due to the shape of its snare which resembles the basilica or the dome-shaped cathedral of the early Roman church. In fact, the orb web is originally constructed as a horizontal orb web structure but is pulled into a dome by the spider. Above and below the domed orb web is an irregular cone-shaped labyrinth that resembles a theridiid web. This web can be mistaken for the sheet web of the filmy dome spider, *Neriene radiata;* however, the dome of the basilica orb-weaver is an orb web with radial and spiral strands.

The basilica orbweaver has a seasonal occurrence in the eastern states, being most common during the months of July and August.

FIGURE 104. A domed web of the Basilica Orbweaver, *Mecynogea lemniscata.*

FIGURE 105. A female Basilica Orbweaver, *Mecynogea lemniscata,* suspended on the underside of her domed web.

When found this spider is most always suspended in its web, which is built in low shrubs and trees. The spider ranges from the District of Columbia south to Florida and west to Colorado (Kaston, 1978).

Size: Length of female 6 to 9 mm; of male 5 to 6.5 mm (Kaston, 1978).

Family Araneidae—*Metepeira labyrinthea* (Hentz)—Labyrinth Orbweaver

Identifying Characteristics: The labyrinth orbweaver is distinguished by having an oval abdomen which has a rich reddish brown ground color transected middorsally by a distinctive white folium. The folium resembles a double-headed arrow pointed toward the anterior end and with a black spot at the tip of the arrow. The carapace is an indistinctive brown color with a slightly lighter ocular region. The legs are light brown proximally and dull yellow distally with dark brown annuli. The labyrinth

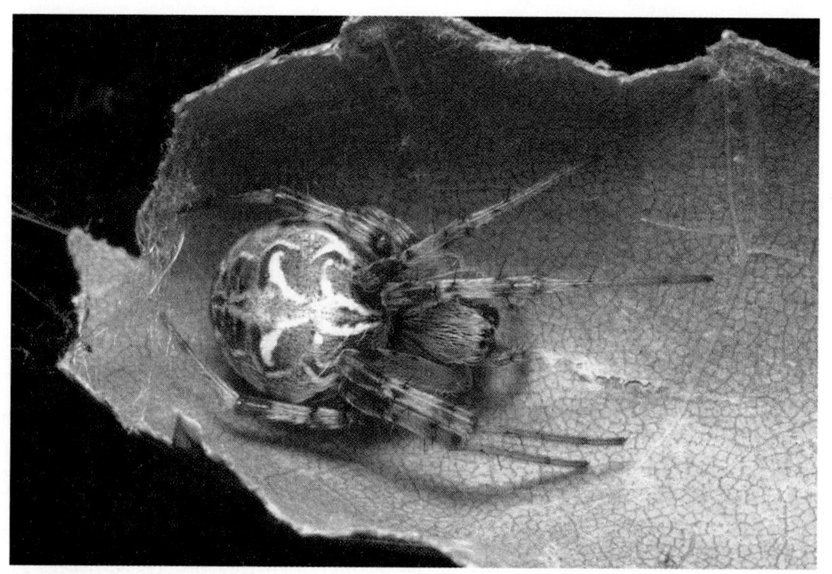

FIGURE 106. A female *Metepeira labyrinthea* sits in her folded leaf retreat.

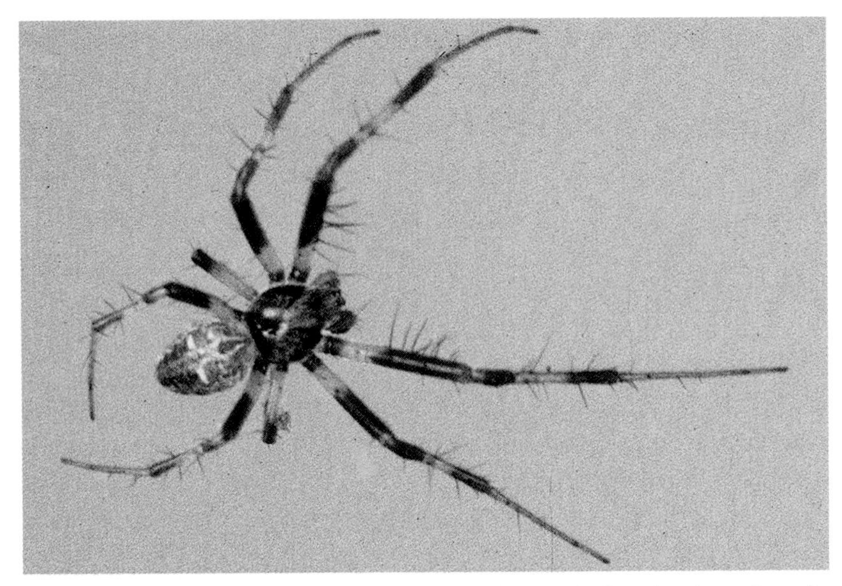

FIGURE 107. Male Labyrinth Orbweaver, *Metepeira labyrinthea,* showing enlarged tarsal palps and the typical abdominal markings for this species.

orbweaver is distinguished from other orbweavers in that the terminal segments of the legs are longer (tarsus plus metatarsus longer than tibia plus patella) and the oval abdomen is held posterior to, rather than above, the cephalothorax.

Ecology and Behavior: As its name implies, this spider has a complex two-part web, consisting of a vertical orb portion, plus an irregular labyrinth component. The labyrinth webbing is constructed behind and above the orb web and is composed of an irregular maze of criss-crossed silken threads, much like the web of a theridiid. The hub of the orb web is an irregular mesh from which one or more traplines extends to a peripheral retreat. The retreat consists of dried leaves sealed together by silken threads which line it. Females mature in early autumn. At this time, male spiders can be found in the webs with the female. In September and October, the female deposits eggs in several tightly-woven egg sacs held together by a single silken string. As evident in the photograph, the egg sacs are lenticular in shape. The egg sacs are constructed near the entrance of the retreat and are securely bound by strong silk to a twig or tree branch. Months later, during winter and early spring, the spider is dead and the webs are often destroyed but the egg sacs can be found still anchored to its twig. The spider is common throughout the U.S. Its range also extends into Canada and west to Alaska.

Size: Length of female 5.5 to 6.2 mm; of male 4 to 4.5 (Kaston, 1978).

Family Araneidae—*Micrathena gracilis* (Walckenaer)—Spined Micrathena

Identifying Characteristics: The females of this very odd spider are easily identified by the presence of a shiny abdomen armed with five pairs of conspicuous black-tipped spines around the margins. The first pair of spines projects forward and is located at the anterior end of the abdomen just behind the pedicel. The second pair is located about midway from the pedicel area to the posterior end of the abdomen, and projects laterally. The remaining three pairs are situated at the posterior end of the abdomen and point backwards at a forty-five degree angle from the midline. There is a great deal of variation in coloration with some individuals having the abdomen almost entirely black, while others have it completely white. In general, the abdomen is whitish with dark spots; the carapace is shiny black with yellowish margins. The sternum is white with brown spots. The legs are largely black, but the joints are yellow beneath.

Males are strikingly different from the females, being much smaller in size and having the abdomen flattened and truncated at the posterior end to form a somewhat elongated rectangle shape which is slightly widest near the middle of its length.

Ecology and Behavior: The orb web of the Spined Micrathena is common in mesic hardwood forests. They are usually constructed from three to

FIGURE 108 A-B. Female (**A**) and male (**B**) *Micrathena gracilis*.

seven feet off the ground and hung between small trees and bushes. The web is very tight, having close concentric circles and radii. Obviously, the web is adapted for capturing small insects. Fitch (1963) found that leafhoppers were the most common insect trapped in the orb snare of this spider. These spiders are exceedingly abundant during the months of July, August and September. The females deposit their egg sacs during September (Moulder, 1992).

Size: Length of female 7.5 to 10 mm; of male 4.5 to 5 mm (Kaston, 1978).

Family Araneidae—*Micrathena mitrata* (Hentz)—White Micrathena

Identifying Characteristics: The females have a hard, glossy white (rarely yellow) abdomen. On the dorsal surface of the anterior abdomen is a black tree-shaped pattern. The trunk and roots of the "tree" are directed anteriorly. A white spot is present in the center of the tree's crown.

FIGURE 109. Female *Micrathena mitrata*

Posteriorly, a black "Mickey Mouse" face pattern is often formed on the white background, the ears projecting anteriorly. The cephalothorax is brownish yellow to black with a gray marginal stripe, and its posterior portion is overlapped by the abdomen. Two pairs of small spines project from the posterior margin of the abdomen. The two spines on each side are aligned over one another so that, when viewed from above, the top spine obscures the lower spine. The legs are translucent gray to yellow with dark markings at the joints. Males are slightly smaller than females, and the spines are only weakly developed.

Ecology and Behavior: This orbweaver, like *M. gracilis,* is very common in mesic hardwood forests which have an understory of bushes and small trees. The tight, compact web is much like that of *M. gracilis* in that it is constructed of close concentric rings and radii. It is about 6 to 8 inches in diameter and is placed from three to seven feet above the forest floor. According to Fitch (1963) these tight webs are perfect for capturing small insects. It seems that they are particularly effective in capturing leafhoppers, the major dietary component of the White Micrathena as well as that of *M. gracilis* and *M. sagittata.* From July through September the webs of all three species of *Micrathena* can be seen in the same area. The White Micrathena matures during August and September and usually disappears toward late October.

Size: Length of female 5 to 5.2 mm; of male 3.8 to 4.0 mm.

Family Araneidae—*Micrathena sagittata* (Walckenaer)—Arrowshaped Micrathena

Identifying Characteristics: The female of this species is perhaps one the most uniquely shaped and beautifully colored of all southern spiders. Its abdomen is shaped much like an arrowhead with the arrow tip at the anterior end of the abdomen. Posteriorly, the abdomen widens and flares outward to form two divergent spines. Another pair of abdominal spines

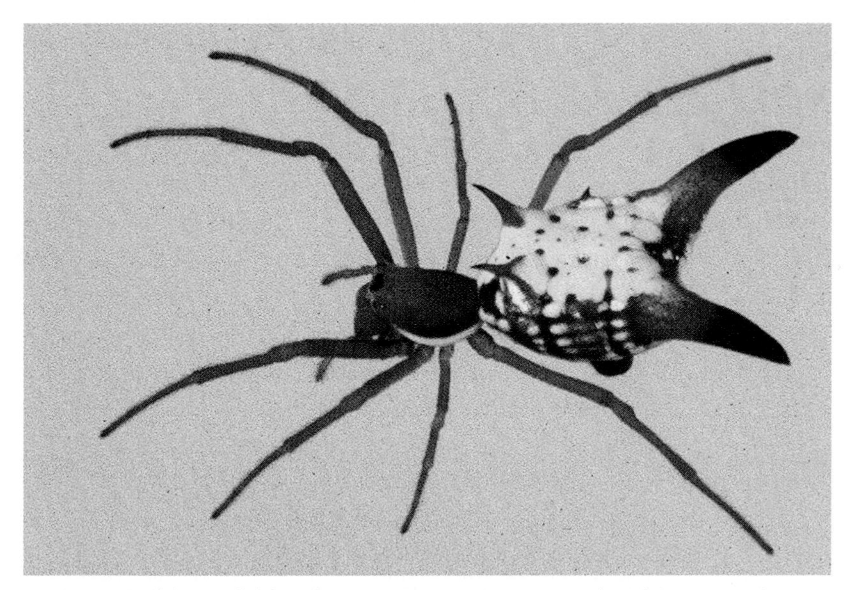

FIGURE 110. A beautiful female *Micrathena sagittata,* armed with its protective spines.

projects dorsally from the anterior end and another projects laterally from the mid region of the abdomen. The three pairs of spines are red at their base and black at their tip. The dorsum of the abdomen is a strikingly bright yellow covered with black dots. Rarely, some specimens have a white abdomen instead. Ventrally, the abdomen bears black bands and yellow spots. The cephalothorax is largely amber colored with a yellow line surrounding its edges. All legs are amber colored.

Males differ from the female in color and size. The male abdomen is widest at the posterior end and it has humps instead of sharp spines. The male abdomen is shiny black to gray brown above with dull white transverse rows and spots at the posterior end. Its cephalothorax is shiny black to brown. Legs I and II are black and Leg III and IV are yellow. Each leg bears a longitudinal black stripe running most of its length.

Ecology and Behavior: The orb web of *M. sagittata* is very tight and compact, being about one foot in diameter. The spider constructs it on low bushes in mesic, deciduous forest. Normally, the vertical web is slightly inclined, with the spider hanging upside down on its downward slope. The web has many radii, and the spiral strands are very closely spaced. A small stabilimentum is often present above the hub of the web. When approached by an intruder, the spider drops out of its web and onto the ground where it conceals itself among leaf litter. According to Fitch (1963) these spiders largely prey upon leafhoppers. From July to September these spiders are quite common. Adults normally die around the time of the first frost in the fall, usually in October.

Size: Length of female 8 to 9 mm; of male 4 to 5 mm (Kaston, 1978).

Family Araneidae—*Neoscona arabesca* (Walckenaer)—Arabesque Orbweaver

Identifying Characteristics: The arabesque spider can be identified by its oval abdomen with three dorsal bilateral pairs of slanted black spots, separated by a median white band. These black spots are slit-like and angled toward the posterior. Each black spot is surrounded by a thin ring of white, lending contrast to the spot. Kaston (1978) distinguishes the male of the species (verified in northern but not southern specimens) by having a curved tibia II. In both sexes there are numerous macrosetae on the ventral surface of tibia II. However, there is variation between the sexes and north-south populations in the distribution of the macrosetae on the tibia. In the female the epigynum lacks lateral bulges and has a short scape. The genitalia are used to positively identify and separate this species from other closely-related *Neoscona* species (Berman and Levi, 1971).

Ecology and Behavior: This spider builds a large vertical orb web in shrubs, small trees, and meadows. The webs are often found in sunny

FIGURE 111. Young female of *Neoscona arabesca.*

and moist situations. This is a nocturnal species. During the daylight hours, if its web is found, the spider may be found nearby just beyond the edge of the web in a retreat formed from a rolled-up leaf. The species is abundant from July until October. The egg sac is lens-shaped, 10 mm in diameter, and contains about 280 eggs (Jackman, 1997). *N. arabesca* is the most widely distributed species of the genus, ranging throughout Canada, the United States, Mexico, and the Caribbean. Nontheless, it constitutes a greater proportion of the *Neoscona* populations in northern regions of the U.S.

Size: Length of female 5 to 12.3 mm; of males 4.2 to 9.2 mm (Kaston, 1978).

Family Araneidae—*Neoscona crucifera* (Lucas)

Identifying Characteristics: *Neoscona crucifera* has long been confused with *N. domicilorum*. Not until 1941 were the two species correctly distinguished (Archer, 1941). As with *N. domicilorum,* the abdomen is triangular to suboval in shape and, on the venter, are two pairs of white spots. There is considerable north-south and east-west variation in size and markings for this species. Without microscopic examination, this spider is distinguished from *N. domicilorum* by its rusty-red coloration and the lack of a distinct dorsal abdominal pattern. Definitive identification of *N. crucifera* must involve a verification of a long scape and two pairs of lateral bulges at the base of the epigynum. In the male, the conductor of the palpus is sigmoid in shape (Berman and Levi, 1971).

Ecology and Behavior: *N. crucifera* often occurs in "hot spots" in that it outnumbers other *Neoscona* species at certain localities. Their large orb

FIGURE 112. Female *Neoscona crucifera* exhibiting the characteristic rusty-red coloration.

webs are more common in open woods and less common in grass. It is often seen around porches and under the eaves of houses, where it constructs its web six or eight feet off the ground (Moulder, 1992). Its range includes all of the southern states, northward into Canada, westward to the Rocky Mountains, and southward into Mexico.

Size: Length of female 8.5 to 19.7 mm; of male 4.5 to 15 mm (Kaston, 1978; Berman and Levi, 1971).

Family Araneidae—*Neoscona domicilorum* (Hentz)

Identifying Characteristics: This orbweaver has a triangular to suboval shaped abdomen that has characteristic markings. The dorsum of the abdomen has a middorsal white band which extends from the pedicel to the region of the spinnerets. Anteriorly, a broad transverse white band

FIGURE 113. Female of *Neoscona domicilorum*.

extends laterally on each side from the middorsal white band to form a distinctive cross-like pattern. In younger specimens there are two to three additional white transverse bands on the abdomen. These are obscured in older specimens such as the one in the above figure. Three pairs of black slit-like spots sit posterior to each transverse white band. These black spots are horizontal in orientation, thereby distinguishing this species from *N. arabesca* which has angled spots. Definitive identification of the species is based upon the presence of a short scape and one pair of bulges on the epigynum (Berman and Levi, 1971). In the male, the terminal apophysis is elongate and straight.

Ecology and Behavior: The webs of *N. domicilorum* are often constructed around buildings. Also, webs may be built in woods which vary in moisture level and openness. We have collected *N. domicilorum* from webs thirty feet above the ground and some spanning one-half the width of a two lane road. This is a nocturnal species. When its web is found during the day, the spider may often be found at the web's edge in its retreat which is formed from a folded leaf. The distribution ranges from Massachusetts west to Indiana and south to Florida and Texas. It is more common in its southern range and is the most frequently encountered *Neoscona.*

Size: Length of female 7.2 to 16.2 mm; of male 6 to 9 mm (Kaston, 1978).

Family Araneidae—*Neoscona pratensis* (Hentz)

Identifying Characteristics: The body coloration of this orbweaver is distinctively different from that of the more commonly encountered species of *Neoscona* (i.e., *arabesca, domicilorum* and *crucifera*). The dorsum of its abdomen has several longitudinal stripes of brown, tan, black and yellow, which extend throughout its length. The middorsal stripe is dark brown. On each side of this stripe is a yellow stripe followed by red, tan, black, yellow and red stripes, respectively. The carapace and legs are amber in

FIGURE 114. Female of *Neoscona pratensis* crawling across the white petals of a summertime flower.

color. The legs bear stout black bristles. The male is colored similarly to the female. A taxonomic treatise of the genus *Neoscona* with species descriptions and keys was published by Berman and Levi (1971).

Ecology and Behavior: This spider most often constructs an orb web among shrubs, low herbs and grasses. Berman and Levi (1971) reported it from salt marshes in New England, New York, Florida, and a bridge in Minnesota. They also collected it by sweep netting upland fields in the George Reserve, Michigan. They also found it in pine-palmetto in Everglades National Park. We have collected it in small shrubs around the salt marshes on Dauphin Island, Mobile County, Alabama. Berman and Levi (1971) felt that this species had a preference for marshes and swamps. Fitch (1963) reported that this species was rarely found in Kansas and the specimens that he collected were all taken from tall-grass in prairie habitats. The first mature specimens are found during May in

Florida and during August in Michigan (Berman and Levi, 1971). The species is found from New England southward through North Carolina to Florida; thence westward to Louisiana and northward to North Dakota (Berman and Levi, 1971).

Size: Length of female 10.2 mm; of male 7.9 mm.

Family Araneidae—*Verrucosa arenata* (Walckenaer)—Triangulate Orbweaver

Identifying Characteristics: The female has a distinctive white and shiny, triangular shield, covering the abdomen. The apex of the triangular shield is directed to the posterior of the abdomen. The triangle is surrounded by highly contrasting purple to pink coloration. The carapace is amber in color. The venter and sternum are dark brown with two light spots midway on the abdomen. The legs are yellow to amber with dark

FIGURE 115. Female of *Verrucosa arenata*.

brown rings. The male is strikingly different from the female in that the abdominal triangular shield is pink and covered with tiny white spots. Three pairs of white tubercles are present on the rear of the abdomen. The cephalothorax and legs are colored as in the female. The venter of the abdomen is purple with the sternum amber. The palps of the male are enlarged and have the appearance of "boxer gloves."

Ecology and Behavior: This distinctive spider is found in eastern hardwood forest from July until the first hard frost usually in October. It is most abundant during August and September and occurs along with related species such as *Micrathena gracilis, Micrathena sagitta,* and *Micrathena mitrata.* The female *V. arenata* is unique among the orbweavers in that it sits at the center of its coarse web with its head pointing upward. In all but dense wooded areas the female reconstructs her web every evening and removes it prior to dawn. The web of the male is not an orb but a mass of irregular strands cast in vegetation at the periphery of the female's web. A courtship occurs over seven to ten days during July and August when the female pays evening visits to the male's web. On the last evening of the courtship, copulation occurs in the male's web. It is found from New York south to Florida and west to Kansas and Texas (Kaston, 1978).

Size: Length of female 8 to 9 mm; of male 5.5 to 6 mm (Kaston, 1978).

Family Araneidae—*Wixia ectypa* (Walckenaer)

Identifying Characteristics: This spider is readily identified by the unique form of the abdomen. It is elevated anteriorly with a bifurcated protuberance forming two "humps" which stick up above and slightly over the head region when at rest. Because the abdomen is raised anteriorly it appears that the cephalothorax is attached by the pedicel to the middle of the abdomen's length when viewed from the side (Figure 116A). The posterior median eyes (PME) are larger than the anterior median eyes, and the PME are located on a raised prominence (Figure 116B). The

FIGURE 116 A-B. Lateral (**A**) and, head-on (**B**) views of the unique "humped-back" spider, *Wixia ectypa.*

spider's general body colors are beige and brown without any notable patterns.

Ecology and Behavior: *Wixia* spins its web at dusk. The web is nearly horizontal and consists of only 7 or 8 radii without any viscid spiral (Kaston, 1948). Little else is known about the biology of the spider. Kaston (1948) considered *Wixia* to be a rare species. Kaston (1948) cited records from Alabama and Florida. This is undoubtedly a species of the southern states.

Size: Length of female 7 to 7.5 mm; of male 6.5 mm (Kaston, 1948).

FAMILY LYCOSIDAE—Wolf Spiders

The wolf spiders are common predators that stalk and chase their prey much like the hunting tactics of wolves. The Greek derivation of "*Lycosa*" means wolf. Wolf spiders run quickly through grass and fallen leaves and can lurk out of sight beneath stones.

A wolf spider is recognized by its characteristic three rows of eyes. The four eyes of the first row are small and situated on the front of the cephalon. The second row of two eyes are the largest eyes and are directed forward. The third row of two large eyes are atop the carapace directed for rear and lateral vision. Lycosid spiders have exception day-light vision for a spider and most have effective night vision.

FIGURE 117. The spiderlings of true wolf spiders cling to their mother's abdomen until they are able to fend for themselves. This female *Rabidosa rabida* is carrying over one hundred spiderlings.

All wolf spiders will carry their egg sacs and tend to their young. Females attach their egg sac to their spinnerets by a bundle of silken threads. When the young begin to hatch, she opens the sac to assist the emergence of the spiderlings. After hatching, dozens of young will ride piggy-back for two to three weeks on the top and sides of the abdomen of their mother.

Lycosidae contains fifteen genera with over 220 species in North America (Roth, 1993). The wolf spiders in the genera *Hogna, Rabidosa,* and *Varacosa* were previously placed in the genus *Lycosa* (Jackman, 1999). The family is widespread throughout most of the U.S. However, many of the species have a specific habitat preference and are restricted in their range to small geographic regions (e.g. *Arctosa sanctaerosae* is limited to sand dunes along the beaches of the Gulf of Mexico in Alabama, Mississippi and the Florida panhandle).

Family Lycosidae—*Arctosa littoralis* (Hentz)

Identifying Characteristics: This wolf spider and other members of the genus *Arctosa* are distinguished from all other spiders by having tarsus I with a dorsobasal bristle which seems to be drawn out into a very thin and fine hair at the end, and which is much longer than the other hairs and trichobothria (Kaston, 1978). The body coloration ranges from mottled to spotted gray to dirty white. The legs are spotted and vary from gray or white to pale beige.

Ecology and Behavior: The pale, mottled body colors make this species well camouflaged against the variegated background of sandy soils. This species is one of about twenty species of *Arctosa*. This species inhabits only beaches, stream banks, and other sandy areas. Although this spider is a voracious predator in its own right, its camouflage serves as protection from larger animals that might otherwise prey on it. This spider usually hides during the daytime underneath logs, driftwood, debris,

FIGURE 118. Female *A. littoralis* displaying its typical coloration to blend in with a sandy substrate.

etc., or in a burrow that the spider digs in the sand. Most of its hunting is done after nightfall as it runs over the sand chasing insects and smaller spiders. This spider can be easily spotted on the sand at night by the use of a headlamp or flashlight held on a level with the collector's eyes. When the flashlight beam hits the spider, the spider's eyes reflect the light in a characteristic blue-green glow. We have collected this spider along sandy banks of rivers, and along the beach sands of the Gulf of Mexico and the Atlantic Ocean. The species ranges throughout the United States and eastern Canada.

Size: Length of female 11 to 15 mm; of male 11 to 15 mm (Kaston, 1978).

Family Lycosidae—*Arctosa sanctaerosae* (Gertsch and Wallace)

Identifying Characteristics: This arctosid is distinguished by being chalk-white or pale orange in color inclusive of every body part: chelicerae, legs, cephalothorax, sternum, and abdomen. On the dorsal surface of the abdomen is a pale gray cardiac mark. The white legs of *A. sanctaerosae* have no dark rings around them, a feature distinguishing this species from other members of *Arctosa*. However, large black setae, referred to as macrosetae, occur on leg I (tibia, femur, tarsus) and leg III (tibia.) The presence of a single dorsal macrosetae on tibia I is taxonomically important in identifying this species. There is very little sexual dimorphism, except for the stout copulatory tubes in the male.

Ecology and Behavior: *Arctosa sanctaerosae* habitat is restricted to the sandy beaches on the northern rim of the Gulf of Mexico. The first

FIGURE 119. Night photograph of a female *A. sanctaerosae* hunting insects on the white beach sands of Dauphin Island, Alabama.

collected specimens came from the beaches of Santa Rosa Island, Pensacola, Florida, thus the specific name. Since then, specimens have been collected from Mississippi to the Florida panhandle (Dondale and Redner, 1983). Adult *A. sanctaerosae* dig burrows with openings of 12 to 15 mm in diameter. Most of the burrows that we have located have been on the secondary dunes of ocean front beaches. Of the burrows that have been excavated, spiders may be 12 to 18 inches deep. On a warm evening, the spiders can be found in the vicinity of their burrow opening, stalking small insect prey. This white spiders are well-camouflaged against the white sand.

Size: Length of female 11 to 12 mm; of male 8 to 13 mm (Gertsch and Wallace, 1934).

Family Lycosidae—*Geolycosa escambiensis* (Wallace)

Identifying Characteristics: *Geolycosa escambiensis* is a burrowing wolf spider. This genus is distinguished by lacking the dorsal spines on the posterior tibiae in females, although they are present on these same regions in males. When viewed from the side, the cephalothorax is very high in the head region and slopes steeply as it approaches the region of the pedicel. The retromargin of the cheliceral fang furrow bears three teeth. Males have a middorsal white stripe on the head bordered by a black stripe on either side. Conversely, the dorsum of the abdomen has a conspicuous black middorsal stripe bordered by a white stripe on either side. While the males have very contrasting black and white stripes, the females are rather plain grayish to bluish-gray in color.

Ecology and Behavior: There are nearly 20 species of *Geolycosa*. Many of these burrowing spiders build a turret around the entrance to their burrow. The burrows are all about one-half inch in diameter and usually about 12 inches in depth. However, some burrows may extend to 36 inches in depth, usually straight down but may be enlarged in the middle

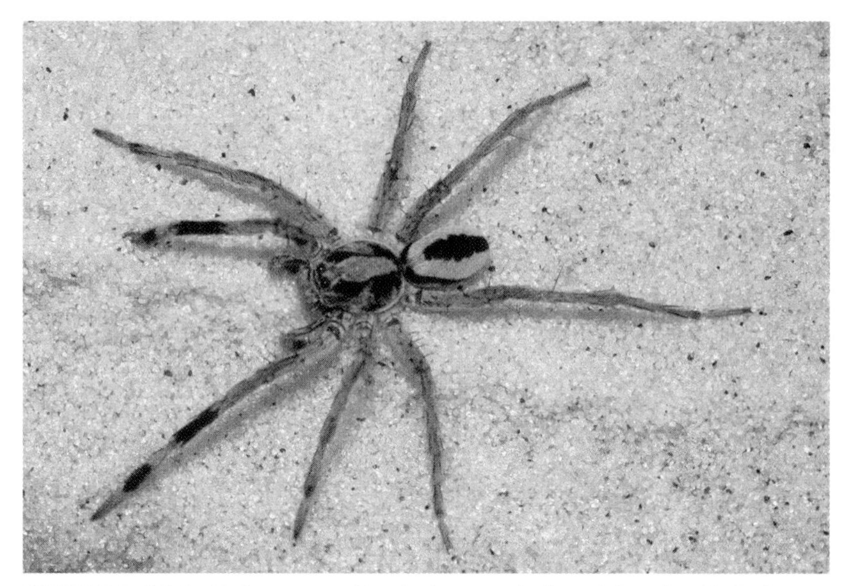

FIGURE 120. Male *Geolycosa escambiensis* photographed at night using flash photography at its sandy beach habitat of Dauphin Island, Alabama.

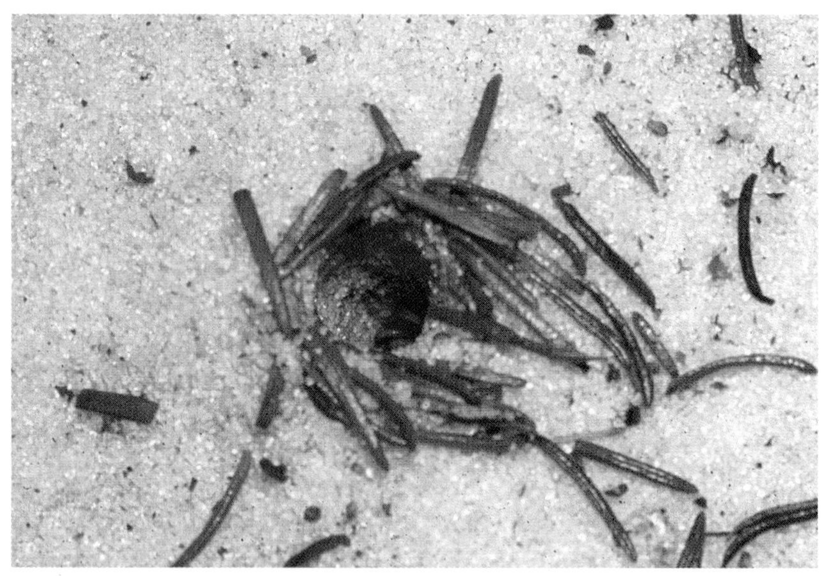

FIGURE 121. Turret of plant materials line the opening into the burrow of *Geolycosa escambiensis* on the sand dunes of the Gulf of Mexico at Dauphin Island, Alabama.

portion of the burrow. The burrow is dug with their chelicerae. The grains of sand are stuck together with silk and thrown out of the burrow. Colored sand from different depths often form differently colored concentric rings around the burrow opening as the spider excavates the burrow. The turret around the burrow opening is often lined with bits of straw, plant debris and pine needles. At night, they usually sit partially inside their burrows in wait of prey. They retreat at the slightest disturbance from human collectors. On sunny days the egg sac may be brought to the surface and warmed in the sunlight. Most specimens must be dug out of their burrows in order to collect them. *G. escambiensis* may or may not have as turret of plant material around their elevated burrow entrance. This species is distributed along the Florida panhandle and northern Gulf of Mexico beaches in areas of white sand and scrub vegetation.

Size: Length of female 18 to 22 mm; of male 14 to 16 mm.

Family Lycosidae—*Geolycosa missouriensis* (Banks)

Identifying Characteristics: This burrowing wolf spider might easily be mistaken for a more common species of lycosid. However, the cephalothorax of *G. missouriensis* is very high in the front and slopes steeply in a straight line to the posterior edge at the pedicel. Unlike other wolf spiders, geolycosid females lack dorsal spines on the posterior tibiae, while males have spines on these same regions. While the basic color pattern previously described for males and females of *G. escambiensis* are essentially the same for *G. missouriensis,* the latter species is darker overall. *G. missouriensis* may be further distinguished from *G. escambiensis* by having the coxa and femur I light below. The venter is light in color in *G. missouriensis* while being black in *G. escambiensis*. A taxonomic revision of the genus *Geolycosa* was given by Wallace (1942).

Ecology and Behavior: Like other geolycosids, this species digs a burrow about one inch in diameter and 12 inches deep in sandy soil. It usually

FIGURE 122. Female *Geolycosa missouriensis* viewed from the side to show the high head region in the vicinity of the eyes but which slopes steeply toward the pedicel. An egg sac is attached to her spinnerets.

constructs a turret composed of bits of vegetation around the burrow opening. In the southern U.S., it has been found relatively common in well-drained and sandy fields. While these spiders almost never leave their burrows, they can be extracted with a small straw which the spider grabs and holds onto when the straw is inserted into its burrow. During April, in Alabama, a female was observed near her burrow, drying her egg sac which was bluish in color and attached to her spinnerets. This geolycosid is the most widely distributed species of the genus (Wallace, 1942). It is distributed from New York south to Alabama and west to South Dakota and Arizona.

Size: Length of female 21 mm; of male 15 to 18 mm (Kaston, 1978).

Family Lycosidae—*Hogna antelucana* (Montgomery)

Identifying Characteristics: This wolf spider can be distinguished from other wolf spiders by the dorsal color pattern. It has a general orange brown color with intermittent patterns of black, gray and sometimes yellow. The carapace has a prominent medial white line, which extends between the median eyes. Behind the eyes this median line widens, becoming widest at the thoracic groove. At the posterior end of the carapace, the median line narrows slightly. Lateral to the median line are wide gray/black bands and white bands at the margins of the carapace. The abdomen has a dark cardiac mark bordered by light bands of white and orange.

Ecology and Behavior: As with other members of the genus, *H. antelucana* is a nocturnal hunter, preying upon insects. It is typically encountered in the evenings but daytime collections from wooded areas are not uncommon.

FIGURE 123. A young *Hogna antelucana* is well camouflaged in the forest litter.

It is a southern species ranging from Florida north to Tennessee and west to California (Kaston, 1978).

Size: Length of female 13.5 to 19 mm; of male 13 to 18 mm (Kaston, 1978).

Family Lycosidae—*Hogna carolinensis* (Walckenaer)

Identifying Characteristics: This is the largest wolf spider in the U.S. Some older authors consider this spider to be in the genus *Lycosa* (Jackman, 1999). The dorsum of this spider is highly variable in color pattern. This variability in body coloration does not allow one to unequivocally distinguish this spider. The carapace of the female is dark brown with interspersed gray hairs, and with a light middorsal stripe extending from the eyes to the pedicel. The carapace is much lighter in males. The abdomen is

FIGURE 124. A female *Hogna carolinensis,* one of our largest wolf spiders.

FIGURE 125. The venter of *Lycosa carolinensis* showing its distinctive black abdomen, sternum, and coxae.

also brown with an indistinctive darker middorsal stripe over the cardiac region. The most distinctive coloration of this spider is seen on the venter where the abdomen, sternum and coxae are all jet black.

Ecology and Behavior: This species, like most lycosids, hunts at night. Most authorities claim that this species constructs its own burrow in the ground in which it hides during the daylight hours (Kaston, 1978). However, we have observed that this wolf spider to be opportunistic in that it will stay in any suitable burrow, whether or not it actually constructs it. These nocturnal hunting spiders actively stalk and avidly feed on crickets, grasshoppers and other insects. Like most lycosids, they mate during late summer and the female may be seen carrying her egg sac beneath her abdomen attached to her spinnerets during September and October. Females carry their young spiderlings on the dorsum of their abdomen for some time after the young hatch.

H. carolinensis is best found by searching grassy fields near woodlands at night by wearing a headlamp on a plane level to one's eyes (Whitcomb et al. 1963). The highly reflective eyes emit a sharp pinpoint of greenish light that can be seen from 40 meters away. *H. carolinensis* is found throughout all of North America.

Size: Length of female 22 to 35 mm; of male 18 to 20 mm (Kaston, 1978).

Family Lycosidae—*Pardosa lapidicina* (Emerton)—Stone Spider

Identifying Characteristics: The ground color of the body is gray to black. Yellow spots are present on the abdomen and cephalothorax but are mostly obscured by dense black hairs which cover the body. The yellow marking on the posterior half of the cephalothorax is nearly bisected by a black line running down the cervical groove. Three or four yellow spots are found around each of the dorsal edges of the cephalothorax. All legs

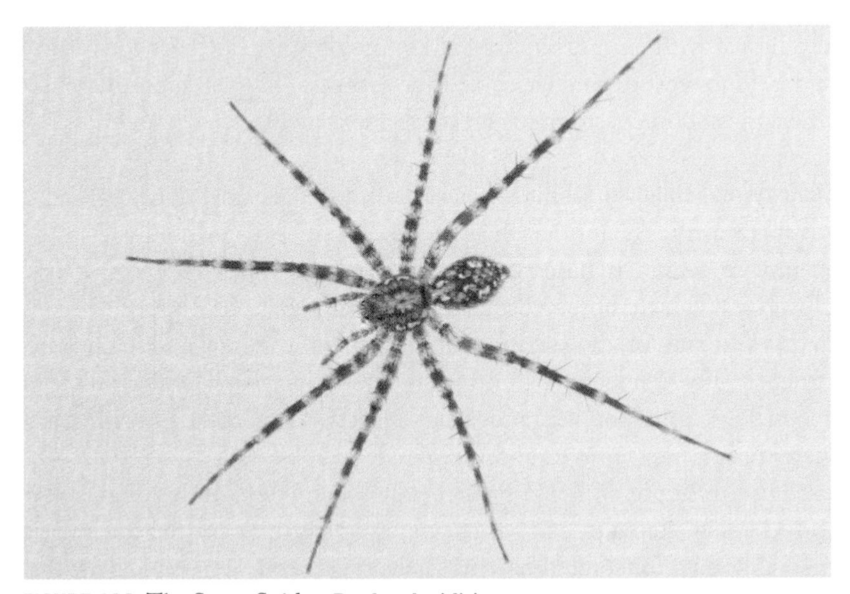

FIGURE 126. The Stone Spider, *Pardosa lapidicina*.

have numerous yellow and black annuli. The dorsum of the abdomen has about three to four yellow spots along each side. The ventral side is a lighter gray than the dorsum. The color varies with the age of the spider, with young, freshly molted individuals having a deeper black color, while the older specimens are grayish (Emerton, 1961).

Ecology and Behavior: Emerton (1961:79) stated, "This spider lives among gray stones in the hottest and driest places from Connecticut to Labrador". Kaston (1978) stated, "These spiders run about swiftly over rocky shores, among stones and clay banks, and among rocks on talus slopes." We have found most of our specimens associated with stones and gravel which occur alongside creeks and rivers. The stone spider often occurs in large numbers in a given habitat. They can run very rapidly and can dart in and out of their numerous hiding places among the rocks. During May many females may be seen carrying egg sacs. The stone spider occurs from New England south to Alabama and west to Texas and Nebraska.

Size: Length of female 7.7 to 9.3 mm; of male 6 to 7 mm (Kaston, 1978)

Family Lycosidae—*Pardosa milvina* (Hentz)—Shore Spider

Identifying Characteristics: This relatively small spider belongs to the genus known as the "Thin-legged Wolf Spiders." It possesses long, thin legs which bear long spines. It has much smaller chelicerae than most other lycosids. The cephalothorax is highest in the head region giving the face vertical sides. The height of the chelicerae is less than the height of the head. This spider has a dark brown carapace with a broad, yellowish, middorsal band. This band is notched about midway between the eyes and the median furrow. The carapace also has yellowish lateral bands encircling its edges. The abdomen is dark brown and covered with hairs. There is a yellowish dagger-shaped cardiac mark followed by a pair of yellowish spots. On the posterior dorsum of the abdomen are paired yellowish spots which meet in the midline to form two to three chevron-like bands. Each of these yellowish spots contains a small black spot.

FIGURE 127. The Shore Spider, *Pardosa milvina.*

Ecology and Behavior: This lycosid is often found in large numbers in bare sandy or muddy shorelines left by receding waters at the edge of a pond or swamp. They are also found along the shores of creeks and other damp areas. They may also be found in leaf litter in thick forested areas. The spiders usually feed on small insects. They move rapidly in darting movements. If alarmed, they travel with great rapidity in a series of long hops. Adults may be found all during the year. Breeding occurs throughout most of the warmer months. The egg sac is carried by the female attached to her spinnerets. Adults may be seen carrying egg sacs from April through September (Fitch, 1963). At least two egg sacs are produced per season. This spider is found from New England and adjacent Canada south to Florida and west to the Rockies (Kaston, 1978).

Size: Length of female 5.2 to 6.2 mm; of male 4 to 4.7 mm (Kaston, 1978).

Lycosidae—*Rabidosa punctulata* (Hentz)—Dotted Wolf Spider

Identifying Characteristics: This handsome spider has a yellowish, cream-colored body which is in striking contrast to the dark brown stripes which course down the carapace and abdomen. The dorsum of the carapace has a pair of dark brown stripes which extend from the eyes posteriorly to the end of the carapace. It also has a middorsal dark stripe which extends the length of the abdomen. A pair of dark stripes extend from the anterior eye row down the front of the chelicerae. The venter of the abdomen is punctuated with distinctive black dots. The sternum is yellow. The large, powerful chelicerae have a boss, scopula, and toothed fang margins. The strong legs are long, tapered and built for running. All legs have spines and the tarsi bear 3 claws. This spider has long been known as *Lycosa punctulata* until recent taxonomic changes (Jackman, 1997; 1999).

FIGURE 128. The Dotted Wolf Spider, *Rabidosa punctulata,* searches for prey among the dead leaves and grasses of a hardwood forest.

FIGURE 129. Ventral view of the Dotted Wolf Spider, *R. punctulata,* showing the distinctively dotted abdomen.

Ecology and Behavior: This spider is found in grassy fields, in open grassy areas near woodlands, and along grass-lined roadside ditches. The spider hunts its prey at night. At dusk, this wolf spider comes out of daytime hiding places and becomes a ferocious, active hunter feeding on grasshoppers, crickets, roaches and other insects. Adults appear in the early summer and are active until the late fall. Mating occurs during late summer and females may be seen carrying their egg cases attached beneath their abdomens to their spinnerets throughout September. It ranges from New England south to Florida and west to the Rockies (Kaston, 1978).

Size: Length of female 11 to 17 mm; of male 13 to 15 mm (Kaston, 1978).

Family Lycosidae—*Rabidosa rabida* (Walckenaer)—Rabid Wolf Spider

Identifying Characteristics: This spider strongly resembles *Rabidosa punctulata* in color and in color patterns. It too, has the ground color of the body yellow, with two longitudinal dark brown stripes running the length of the cephalothorax, and a single dark longitudinal stripe extending the length of the abdomen. The differences are that the rabid wolf spider has the median brown stripe on the abdomen disrupted by wavy margins and by enclosing pairs of white spots which form about 5 or 6 rows down the midline. *R. punctulata* has this stripe with straight margins and does not enclose pairs of white dots. Additionally, *R. rabidosa* may be further distinguished from that spider by having, in the male, leg I dark brown or black, and the venter is not spotted.

Ecology and Behavior: The habitat and habits are the same as those of *R. punctulata.* The spider lives in tall grasses in fields and along the edges

FIGURE 130. Female Rabid Wolf Spider, *Rabidosa rabida.*

of woodlands where it is a nocturnal hunter. It does not construct a snare, but lies in wait to pounce upon its prey, usually insects. It is one of the most common, most widely distributed and best known of North American wolf spiders. Oftentimes, it is found in lawns and backyard lots and may wander into residential swimming pools. It often can be seen in late Fall, usually September, with its egg case attached to its venter and held by the spinnerets. After emerging from the egg case, the young climb onto the abdomen of the mother where she carries them for several days. The egg sacs are from 7 to 10 mm in diameter and contain from 168 to 365 eggs (Jackman, 1997). *R. rabida* is distributed from New England to Florida and west to Nebraska and Oklahoma (Kaston, 1978).

Size: Length of female 16 to 21 mm; of male 11 to 12 mm (Kaston, 1978).

Family Lycosidae—*Schizocosa avida* (Walckenaer)—Lance Wolf Spider

Identifying Characteristics: This is a relatively large wolf spider with very distinct markings. A conspicuous middorsal light stripe extends from the eyes to the end of the cephalothorax. Along the sides of the cephalothorax are light stripes about the same color and width as the middorsal stripe. These stripes extend through an otherwise dark brown to gray cephalothorax. Along the anterior and dorsal portion of the brown to grayish abdomen are two light stripes that unite at about the middle of the abdominal dorsum to continue as a single, light stripe to the area of the spinnerets. These two light stripes form a distinctive "V" pattern on the anterior and dorsal portion of the abdomen, and enclose a dark brown lanceolate mark. The legs vary from yellowish brown to gray.

Ecology and Behavior: This relatively large wolf spider is commonly seen in open fields and pastures. Like most lycosids, it is a wandering hunter of prey. In the south, we have found it commonly in pine forests among the brownish pine needles on the forest floor. Fitch (1963) reported

FIGURE 131. Female *Schizocosa avida,* "rescued" from a neighborhood swimming pool.

being bitten on the finger by an adult *S. avida*. He stated ". . . there was a sharp prick from the fangs and a sensation of numbness that lasted only momentarily." Kaston (1948) found that individuals mature in late May and males are found until the end of June. Females live until the fall and construct egg sacs from June to August. The female carries her egg sacs which are 7 to 8 mm in diameter and contain from 119 to 201 eggs (Kaston,1948). We have found many specimens that have wandered into residential swimming pools and have been either drowned or trapped in pool filters. It ranges throughout the U.S. and Canada (Kaston, 1978).

Size: Length of female 10 to 15 mm; of male 8 to 11 mm (Kaston, 1978).

Family Lycosidae—*Schizocosa crassipes* (Walckenaer)—Brush-Legged Spider

Identifying Characteristics: The carapace has a wide, middorsal light beige stripe, bordered by marginal brown stripes. The middorsum of the abdomen has a wide, light area somewhat darker than the carapace and bordered by dark stripes with light and dark spots intermingled. The legs are yellowish with black annuli. The males resemble the females but have a conspicuous brush of black hairs on tibia I, a smaller brush on patella I, and almost none on metatarsus I.

Ecology and Behavior: This medium-sized wolf spider lives in woodland habitats among the leaf litter of the forest floor. Its general body coloration of brown, beige, and black with spots and stripes, matches the coloration of dead leaves. Like most lycosids, this spider is an active wandering hunter. Fitch (1963) found that this was the most abundant

FIGURE 132. Male *Schizocosa crassipes* with distinctive black hair brushes on Leg I.

spider of its size group at a study site in Kansas. He found that the species was also active during midwinter. He even reported them running through grass and patches of snow during January. Kaston (1948) also found that this species in Connecticut overwinters in the antepenultimate instar, matures in late April or May, and produces egg sacs in July and August. Fitch (1963) reported large numbers of males in Kansas in early June, which he believes to be their breeding season. This species is distributed from New England south to Georgia and west to Oklahoma and Nebraska (Kaston, 1978).

Size: Length of female 6 to 8 mm; of male 6 to 8 mm (Kaston, 1978).

FAMILY PISAURIDAE—Nursery Web Spiders

This family contains medium to large spiders that superficially resemble the wolf spiders. Like the wolf spiders, the pisaurids wander about in search of prey, rarely constructing snares. Furthermore, the pisaurids are strongly maternalistic. The females carry and protect their egg sacs with the chelicerae and palps. Before the young emerge from the egg sac, the female builds a nursery web around the egg sac in the branches of a tree or bush.

Distinct from the wolf spiders, the adult pisaurid generally rests in a flattened position with legs extended. When disturbed or enticed by prey, they can move with exceptional quickness. This group is also characterized by having a broad abdomen that tapers at the posterior end.

Similar to the wolf spiders, the pisaurid spiders have their eyes arranged in three rows. However, in the pisaurids the ocular quadrangle is a trapezoid shape because the posterior lateral eyes are distantly separated. Furthermore, the eyes of the second and third row of the pisaurids are nearly the same size, but this is not the case among the wolf spiders. Similar to wolf spiders, the tarsi of the pisaurid spiders have three claws. However, in pisaurids the median claw has two or three teeth, rather than one or none among the wolf spiders.

FIGURE 133. A female *Pisaurina* sp. surrounds her nursery web containing hundreds of tiny spiderlings.

Pisauridae consists of three genera and 13 species (Roth, 1993). The two largest genera include *Dolomedes* (fishing spiders) and *Pisaurina* (nursery web spiders). Excellent taxonomic treatises and keys to this group have been presented by Carico (1973a; 1973b; 1993) and Dondale and Redner (1990).

Family Pisauridae—*Dolomedes albineus* (Hentz)

Identifying Characteristics: This is a strikingly handsome spider. The ground color of the carapace is tan, but this color is obscured by a dense covering of white hairs. Thus, a bold white carapace contrasts against the legs and abdomen and immediately draws one's attention when first seeing this spider. The cephalic portion of the carapace is well elevated above the thoracic carapace. The abdomen and legs also have dark brown markings.

FIGURE 134. A female *Dolomedes albineus* displaying her distinctive white carapace and typical abdominal markings.

Two patches of white hairs delineate the anterior portion of the abdomen. Five W-shaped, dark brown, transverse lines cross the abdomen. Adjacent W-shaped lines are separated by dirty white bands. White dots occur at the lateral arms and at the elbows of the W's (Carico, 1973).

Ecology and Behavior: According to Carico (1973), this large spider is most often found resting on the trunks of emergent trees in swamps, ponds and slow-moving streams. It has the habit of positioning its body with its head in a downward position facing the water or ground. When disturbed, it either rushes into the water or skitters to the opposite side of the tree (Carico, 1973). Very little is known about its life history.

Size: Length of female 23 mm; of male 18 mm.

Family Pisauridae—*Dolomedes scriptus* (Hentz)

Identifying Characteristics: *D. scriptus* superficially resembles *D. tenebrosus,* a species with overlapping habitat and range. However, differences in the structure of their genitalia distinguish them (Carico, 1973). *D. scriptus* can usually be distinguished from most members of *Dolomedes* by having transverse, alternating, black and white W-shaped marks on the posterior half of the abdomen. Anteriorly, on the carapace the area surrounding the eyes is dark. The clypeus has a transverse band of light hairs interrupted by irregular dark mottled pigment. A pair of triangular spots mark the anterior rim of the thoracic groove. A narrow, medial, light band extends posteriorly from between the thoracic triangular marks and surrounds the thoracic groove. The dorsum of the abdomen has a light gray, lanceolate cardiac area. Coloration is similar in both sexes, except that the legs of the female are annulate and those of the male are not.

FIGURE 135. Male *D. scriptus* rests on a tree trunk at the edge of a small stream as it stalks a tiny mosquitofish.

Ecology and Behavior: This species is associated with moderate to fast-moving streams where they are found among rocks and boulders at the edge of the stream (Carico, 1973). Sometimes they can be found in the water or among piles of dead sticks, leaves and debris washed up on the bank. The mottled grayish color camouflages the spider among the gray rocks, weathered wood, and sticks. Moulder (1992) observed that this species, when hunting, will rest its front three pairs of legs on the surface film of the water and its fourth pair on the shore. He found that aquatic insects and occasionally very small fishes are captured for food. The species is found from southern Canada southward throughout most of eastern and central U.S. (Carico, 1973). Wherever *D. scriptus* is found, a similar species, *D. vittatus,* is found as well.

Size: Length of female 17 to 24 mm; of male 13 to 16 mm (Kaston, 1978).

FIGURE 136. Female *Dolomedes scriptus* resting on a tree trunk.

Family Pisauridae—*Dolomedes tenebrosus* (Hentz)

Identifying Characteristics: The general coloration of this fishing spider is gray to brownish with distinctive dark gray markings especially on the abdomen. The carapace is dark gray in the middle becoming lighter toward the eyes. Across the posterior half of the abdomen are four transverse dark W-shaped stripes. These transverse stripes each have a contrasting light border on their posterior edge. The venter of the cephalothorax and legs are light colored, without distinctive markings, with the abdomen being slightly darker. The legs are marked with light and dark rings and are studded with long dark spines. The sternum is gray at the edges with a median lanceolate mark. Males have a more slender body, longer legs, and are more strongly marked.

Ecology and Behavior: This is a large, robust fishing spider found around bridges, rip-rap, rocks, low bushes and tree trunks near permanent

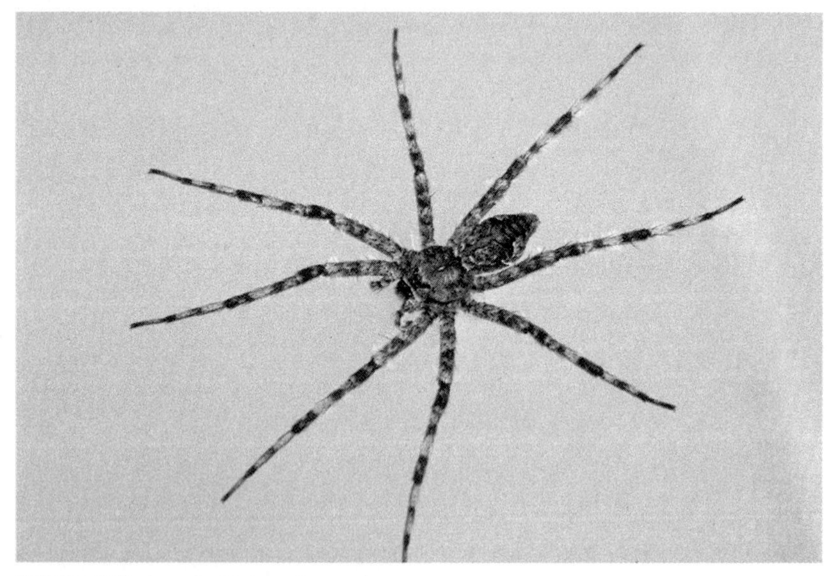

FIGURE 137. A male *Dolomedes tenebrosus.*

bodies of water. Carico (1973) collected *D. tenebrosus* most often from the vertical trunks of trees and other emergent objects from slow-moving ponds, streams and swamps. They can run onto and over the surface of the water with great agility (Carico, 1973). They can remain afloat due to special hairs on the ventral surface of the body that are coated with a water repellent substance. Rarely, the species may be found in dry woodlands far from bodies of water. This spider is most active at night. The female characteristically carries her egg sac in her chelicerae until the young hatch and start to emerge. She will then build a large nursery web in the bushes around the egg sac and young. This web serves as effective protection for the young after hatching. *D. tenebrosus* is distributed from southern Canada throughout most of the eastern United States (Carico, 1973).

Size: Length of female 15 to 26 mm; of male 7 to 13 mm (Kaston, 1978).

Family Pisauridae—*Dolomedes triton* (Walckenaer)—Six-Dotted Dolomedes

Identifying Characteristics: This species was long known under the name *Dolomedes sexpunctatus* Hentz. The color of this distinctive fishing spider is usually a dark green-brown except a contrasting white band on each side extending the whole length of the body, and two rows of white spots on the dorsum of the abdomen. The sternum is marked with six large dark dots, three on each side near the coxae. According to Comstock (1912) the six dots on the sternum probably suggested the earlier name "*sexpunctatus*." The male and female genitalia serve to further distinguish this species from other members of the genus *Dolomedes* (Carico, 1973). The male has a spinous hump on the ventral side of the femur on leg IV (Kaston, 1978).

Ecology and Behavior: Of all the species of its genus, *Dolomedes triton* is the one that displays an exclusive preference for standing-water habitats

FIGURE 138. Male of *Dolomedes triton*.

such as lakes, ponds, or pools in a slow-moving stream (Carico, 1973). Usually, the spider can be found among emergent aquatic vegetation at the water's edge. They position their head toward the water, or with the anterior legs on the water while the posterior legs rest on the vegetation. When frightened, they dive beneath the surface of the water and hide beneath floating leaves and debris. They have been observed to eat adult damsel flies and other water-related insects (Carico, 1973). Egg sacs are laid during April in southern states and June in the northern states. Like other members of this family, this species constructs nursery webs into which they place their egg sacs. The mother remains outside the nursery nearby to protect the young. This is the most widely distributed *Dolomedes*. It is found across much of Canada to southern Alaska, southward throughout most of the U.S., to the Yucatan Peninsula and into Chiapas, Mexico (Carico, 1973).

Size: Length of female 17 to 20 mm; of male 9 to 13 mm (Kaston, 1978).

Family Pisauridae—*Dolomedes vittatus* (Walckenaer)

Identifying Characteristics: This is a large, deep chocolate brown fishing spider that has several distinctive dorsal markings. Sexual dimorphism in color patterns is so great that early naturalists described the two sexes as different species. For both sexes the carapace is dark brown from the eyes to the pedicel. In the female, more so than the male, a thin white band crosses the clypeus and completely encircles the carapace laterally (Carico, 1973). More evident in the male is a pair of distinctive black triangular spots just anterior to the thoracic groove. Several dark lines radiate outward from the thoracic groove and end abruptly at the edge of the central disc. The dark-brown abdomen is encircled at its lateral margins by a light band. In the cardiac region, a pale midline mark extends posteriorly to near the middle of the dorsum. On the dorsum of the abdomen in the female are two pairs of small white spots in the basal half and three pairs of larger white spots in the apical half. Transverse

FIGURE 139. A male *Dolomedes vittatus*.

dark lines join the pairs of spots. Unlike the male, the sides are dark brown. The male has a patch of stiff spines on femur IV.

Ecology and Behavior: According to Carico (1973), *D. vittatus* and *D. scriptus* occur together at many sites. However, *D. vittatus* is more commonly found in small, well-covered streams, while *D. scriptus* is more common in larger, open streams. Carico hypothesized that the darker brown colors in *D. vittatus* might be protective because they inhabit shaded areas of streams around piles of woody debris, around rocks and boulders, and on trunks of trees. This species, like other pisaurids, constructs a nursery web. Egg sacs are deposited in nursery webs from late summer to early fall. Each nursery contains 1000 to 1500 eggs or spiderlings (Carico, 1973). This species is found from southern Canada (Ontario) southward throughout most of the eastern U.S.

Size: Length of female 19 to 28 mm; of male about the same (Kaston, 1978).

Family Pisauridae—*Pisaurina mira* (Walckenaer)—Nursery Web Spider

Identifying Characteristics: This is one of our most boldly and uniquely marked spiders. The ground color of the body varies from yellow to light brown with a darker brown broad median dorsal stripe that extends from the eyes through the cephalothorax and is continued to the posterior end of the abdomen. On the abdomen, the edges of the broad median stripe are scalloped and margined with white. The eyes are ringed with black. The chelicerae are reddish brown. The eight eyes are nearly of equal size. The anterior eye row is straight, and the posterior eye row is so strongly recurved that it seems to form two rows.

Ecology and Behavior: This spider is often found sitting in the tops of small herbaceous plants, grasses and shrubs. It is often associated with

FIGURE 140. Female *Pisaurina mira* exhibiting her distinctive broad dark stripe that extends from the eyes posteriorly to the end of the abdomen.

fields of tall grasses, meadows, and moist, open woodlands. It actively hunts for a large part of the day and for the remainder of the day it sits quietly. It has no permanent home but roams about searching for prey. During courtship, the male will present the female with a fly. If the female accepts the offering, she will feed on it during the mating act. At first, the female carries the egg sac around in her chelicerae for a short time. Then, she will build a nursery web in weeds or low bushes. She stays with the nursery and guards the spiderlings after they hatch for a week or so until they leave the nest area.

Size: Length of female 12.5 to 16.5 mm; of male 10.5 to 15 mm (Kaston, 1978).

Family Pisauridae—*Pisaurina undulata* (Keyserling)

Identifying Characteristics: This spider resembles *Tibellus oblongus* of the Family Philodromidae. Like that spider, *Pisaurina* has the cephalothorax longer than wide and possesses an elongate abdomen. Additionally, *Pisaurina,* like *Tibellus,* has a dull yellow-beige cephalothorax with a pair of black lines extending down the midline. The abdomen is similarly colored with a dusky line extending down the middorsum. This line is a continuation of the one on the cephalothorax. The abdominal line narrows posteriorly and the edges of the line are gently scalloped. The legs are light yellow-brown, a darker shade more distally. The eyes appear to be arranged in four rows because the anterior row of eyes is very strongly procurved and the posterior row of eyes is strongly recurved. Each cheliceral fang furrow bears three teeth along its retromargin. Tibia I and II bear five pairs of ventral spines that are very long and overlapping. Metatarsi I and II have four pairs of very long ventral spines. The ster-

FIGURE 141. Male *Pisaurina undulata.*

num is longer than wide and elongated to a posterior point that extends between coxae IV.

Ecology and Behavior: *Pisaurina undulata* is nearly always found close to the edges of ponds and streams. It is well camouflaged in the emergent vegetation and grasses along the shoreline. We have collected it most often by sweep netting the grasses along streams and ponds. *P. undulata* is essentially a southern species. It is found from North Carolina south to Florida and west to Louisiana (Kaston, 1978).

Size: Length of female 11 to 19 mm; of male 9 to 14 mm (Kaston, 1978).

FAMILY OXYOPIDAE—Lynx Spiders

Lynx spiders get their name because of their aggressive chase of prey, which in many ways resemble that of the lynx, or bobcat. They hide in the tops of tall grasses and shrubs, superbly camouflaged from their prey. When insects come within reach to feed on flowers, the lynx spider surges, in an explosive leap, to grasp the unsuspecting and bewildered prey. They can leap from branch to branch in pursuit of prey much like jumping spiders.

Their eight eyes are arranged in a hexagonal pattern such that the first row is strongly recurved and the posterior row is strongly procurved. The curvature of the rows of eyes is so prominent that there appear to be four rows. The anterior row of eyes is much smaller. All eyes are dark. The legs are especially long and bear long spiny bristles. The tarsi bear two claws and two rows of trichobothria. Posteriorly, the abdomen tapers to a point.

Oxyopidae is a small family of three genera and 20 species in North America (Roth, 1993). This family has been studied taxonomically by Brady (1964; 1975) and Dondale and Redner (1990). *Oxyopes salticus* and *Peucetia viridans* are among the most commonly encountered species and are presented here to represent the family.

FIGURE 142. Face-to-face with a female Green Lynx Spider.

Family Oxyopidae—***Oxyopes salticus*** (Hentz)—Striped Lynx Spider

Identifying Characteristics: The ground color of the cephalothorax and abdomen is yellow. Contrasting with the yellow color of the carapace are four longitudinal brownish lines extending from the posterior eye row to the thoracic declivity. A pair of thin black lines are present on the clypeus and front of the chelicerae. The dorsum of the abdomen is pale yellow with a dark cardiac mark and a pair of lines which taper posteriorly. Dark stripes are present along the sides of the abdomen. On the venter, the sternum is yellow while the bases of the coxae are black. The abdominal venter is pale yellow with a wide median dark band. The femora of legs I, II and III each have a thin, longitudinal black line beneath. The coloration and markings of the male are similar to those of the female except that the male has the abdomen entirely gray or black, dorsally and ventrally, and with scales that give it a metallic iridescence. The anterior row of four eyes is strongly recurved while the posterior row of four eyes is

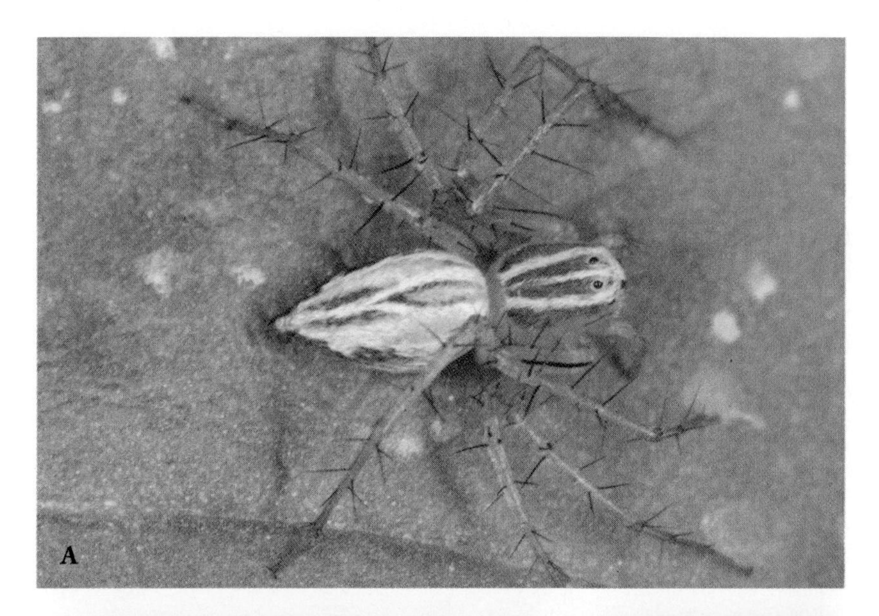

FIGURE 143. A-B. (**A**) Female; and (**B**) brilliantly iridescent male of *Oxyopes salticus*.

strongly procurved. The face is vertical, and the carapace is highly arched. The elongate abdomen tapers to a point posteriorly.

Ecology and Behavior: This spider is abundant in tall grasses during the summer months. This species is a very active, fast-moving spider that runs and jumps among vegetation. According to Fitch (1963), this spider is "scansorial and saltatorial, climbing rapidly and jerkily among stems and leaves, and jumping from time to time." The spider makes no web but prefers to wander about stalking its prey and pouncing upon it. Adults have been observed mating during June (Fitch, 1963). We have collected this species by using a sweep net in tall broomsedge grasses during October. It is distributed throughout the U.S. (Kaston, 1978).

Size: Length of female 5.7 to 6.7 mm; of male 4 to 4.5 mm (Kaston, 1978).

Family Oxyopidae—*Peucetia viridans* (Hentz)—Green Lynx Spider

Identifying Characteristics: *P. viridans* is easily recognized by its bright green body with red spots in the eye region. The red spots may also be over other parts of the body, varying in number and size. The legs are light green to yellow, covered with numerous black spots and protected with long black spines. The green abdomen exhibits three pairs of white chevron-shaped bands that point anteriorly. The abdomen usually tapers to a point posteriorly. The retromargin of the chelicerae fang furrow is without teeth.

Ecology and Behavior: *P. viridans* build no webs, snares or retreats. They live among the tops of low bushes and herbaceous vegetation on which they hunt prey and fasten their egg sacs. These spiders can run rapidly and jump with surprising grace. They prefer to hide among flower heads such as goldenrods and the blooms of other wildflowers. Here, they

FIGURE 144. Camouflaged Green Lynx Spider, *Peucetia viridans*, patiently awaits the arrival of an unsuspecting insect which will provide its next meal.

blend well with the bright green foliage. According to Jackman (1997), bees can often constitute over 20% of their diet. They also prey upon sphecid and vespid wasps and lepidopteran larvae. The spider matures in the summer months and mates in the fall. Their egg sacs are from 1.2 to 2.5 cm in diameter and contain from 100 to 500 eggs. The females guard the egg sac with profound tenacity. The female can spit venom from her fangs when disturbed while protecting her egg sac (Fink, 1984). *P. viridans* is distributed from Virginia south to Florida and west to California (Kaston, 1978).

Size: Length of the female 14 to 16 mm; of male 12 to 13 mm (Kaston, 1978).

FAMILY CTENIDAE—False Wolf Spiders (or Wandering Spiders)

This is a small family of moderately large, hairy, wandering spiders that resemble wolf spiders. The adult females of some species may reach nearly two inches in length. Like the wolf spiders, they prowl the forest floor and foliage in search of prey, and most species are nocturnal. The ctenid spiders have two eyes that make up the first row of eyes, four eyes in the second row, and two in the third row (2–4–2 arrangement). The distinctive eye arrangement distinguishes ctenids from the true wolf spiders (4–2–2 arrangement), as well as from all other spider families. Additionally, the ctenid spiders have only two claws with teeth on each tarsus. Very little is known about the habits of the ctenid spiders. Ctenidae contains three genera and five species. All four species are restricted in their distribution to the southern U.S. and southwest Texas and near Arizona in Sonora (Roth, 1993).

FIGURE 145. The false wolf spider, *Anahita animosa,* climbs over mosses in a moist hardwood forest.

Family Ctenidae—*Anahita animosa* (Walckenaer)

Identifying Characteristics: The cephalothorax, abdomen, and legs are yellowish to orange-brown. The cephalothorax bears a lighter median, longitudinal band that has wavy margins. This band extends from the second row of eyes posteriorly to the end of the cephalothorax. The band on the abdomen breaks up into a double row of white dots extending its length. The legs are long, spiny and covered with black spots. *Anahita animosa* may be recognized by having the ctenid eye arrangement and two claws per tarsus. It possesses three teeth on the retromargin of the cheliceral fang furrow, a feature which distinguishes it from *Ctenus hibernalis,* a closely-related false wolf spider which has four teeth in this same region.

Ecology and Behavior: This false wolf spider may be found in the warmer summer and fall months wandering the deciduous forest floor among

FIGURE 146. A female *Anahita animosa.*

the dead leaves. During daylight hours, we have collected this species during late August through October by raking leaves in hardwood forests. This disturbs both *Anahita animosa* and the closely-related *Ctenus hibernalis,* causing them to scurry about in search of other hiding places among the leaves. At night, head lamps may be used to find these woodland spiders as they scour the forest floor in search of prey. We have found this spider to be very common in the woodlands of northern Alabama. Little is known of the habits of this false wolf spider. *A. animosa* is found in the southern and central U.S. (Kaston, 1978).

Size: Length of female 7 to 9 mm; of male 6 to 8 mm (Kaston, 1978).

Family Ctenidae—*Ctenus hibernalis* (Hentz)

Identifying Characteristics: The cephalothorax is strikingly bicolored. The sides are dark brownish to orange-brown. Along the midline of the carapace is a contrasting light colored band that extends from the eyes to the region of the pedicel. The band varies in color from yellowish to dirty-white to beige. The abdomen is dark brown to orange brown with a lighter contrasting yellowish midline stripe. This stripe has strongly serrated margins. The legs are light brown in color and are armed with spines. The ventral surface of tibia I has five pairs of spines.

Ecology and Behavior: *Ctenus hibernalis* may be found during daylight or evening hours as it wanders on the forest floor in search of prey. It has a southern range that is limited to Alabama, Mississippi, and Florida (Kaston, 1978). Within this range, most will survive all but the harshest winters. We have collected many individuals that did not seem to fear their collectors. They will rest on a person's shoulder or walk casually along one's arm. However, we suggest some caution, especially to children. The venom from its bite can cause the temporary eruption of a welt that may persist for half an hour or more. The ctenid spiders make no web for a dwelling, but may take up residence in an unoccupied

FIGURE 147. Female *Ctenus hibernalis* displaying its distinctive broad, yellow middorsal stripe that extends the length of its cephalothorax and abdomen.

burrow of some other invertebrate. Egg sacs are deposited by the female in the forest vegetation. Different from the true wolf spiders, the adults neither carry their egg sacs by attaching them to the spinnerets, nor do they provide maternal care of young.

Size: Length of female 16 to 18 mm; of male 13 mm (Kaston, 1978).

FAMILY AGELENIDAE—Funnel Weavers

The spiders of the Family Agelenidae spin sheet-like webs in the grasses of pastures, leaf litter and weeds of the forest floor, and in the corners of fences and barns. The web is usually provided with a tubular retreat in which the spider is found sitting. The web in many species has the appearance of a funnel which is the basis for its common family name "funnel weavers." Agelenid spiders are sedentary spiders that rarely emerge from their tubular retreats and are not found wandering about.

When the spider does emerge from its retreat it is to seize prey and then to retreat again. Most distinctive of the family are the long posterior spinnerets that extend noticably beyond the abdomen. Their eyes are in two rows and about the same size. Some species have light colored eyes, some dark colored eyes, and some species have a combination. The family contains 24 genera and an excess of 283 species in North America. Roth (1993) has estimated that an additional 133 species are undescribed.

Family Agelenidae—*Agelenopsis naevia* (Walckenaer)—Grass Spider

Identifying Characteristics: Taxonomically, *Agelenopsis* is a very complex genus with perhaps up to 20 valid species. Nearly all of these species share similar characteristics and are distinguished only by their genitalia. *A. naevia* is the most commonly encountered and largest of the "grass spiders." The spider varies in coloration from pale yellow with gray

FIGURE 148. *Agelenopsis naevia* sitting on its sheet web.

markings to reddish brown with black markings. The cephalothorax has two longitudinal gray stripes extending posteriorly from the lateral eyes, and it is bordered by a black line. The abdomen is gray to black along the sides with a light brown to brownish red middorsal stripe, bordered by light colored spots. There may be a V-shaped mark on the sternum, but if present it is not as distinct as the V-shaped mark on the closely-related *A. pennsylvanica. A. naevia* has a broad medium band on the venter of the abdomen. The cephalothorax and abdomen are covered with fine gray hairs. The posterior spinnerets are very long, with the apical segment twice as long as the basal segment. The legs are long and marked by dark rings at the ends of the joints and lighter rings in the middle of the femora and tibia. The legs bear numerous spines and trichobothria: two rows on the tibia and one row on the metatarsi and tarsi. There are three tarsal claws.

Ecology and Behavior: The distinctive webs of this grass spider may be seen abundantly from June to September in grass, around stone fences,

FIGURE 149. A young male *Agelenopsis naevia* exhibiting its distinctively long posterior spinnerets.

in shrubbery, and in the corners of buildings. These funnel web weavers construct platform webs with a funnel-like or tubular retreat in which the spider hides. The web is not sticky, serving only as a landing area where insects may land. Above the horizontal platform, the spider may construct an irregular silken labyrinth. The labyrinth serves to deflect flying insects out of the air, causing them to fall upon the horizontal web. The spider then pounces upon the prey with lightning speed. The web may reach up to three feet in width. Sexual maturity is reached during late August. Eggs are laid in a white, flat, elliptical cocoon, 8 to 9 mm in diameter, which is attached to the undersides of loose bark. The female stays near the egg sac until the first frost when she usually dies. *A. naevia* is commonly found from New England and adjacent Canada south to Florida and west to Kansas and Texas (Kaston, 1978).

Size: Length of female 16 to 20 mm; of male 13 to 18 mm (Kaston, 1978).

Family Agelenidae—*Agelenopsis pennsylvanica* (C.L. Koch)

Identifying Characteristics: *A. pennsylvanica* is very similar to *A. naevia* with respect to markings, morphology, web construction, behavior and ecology. Both species are often found in similar habitats. However, *A. pennsylvanica* is less frequently encountered. *A. pennsylvanica* is preliminarily distinguished by the broader and more rounded carapace. On the carapace, the pair of longitudinal dark bands is broader than in *A. naevia*. The abdomen of *A. pennsylvanica* is darker laterally and lighter along the midline. Within the light, median band is a dark cardiac mark, anteriorly, and four lighter chevrons, posteriorly. The segments of each leg are marked with dark rings, especially the tibia and metatarsus. The identity of *A. pennsylvanica* can be verified by its ventral markings: a V-shaped mark on the sternum and a broad gray band on the underside of the abdomen (Kaston, 1978).

Ecology and Behavior: As a grass spider, *A. pennsylvanica* constructs sheet-like funnel webs supported by tall grasses of fields, ground

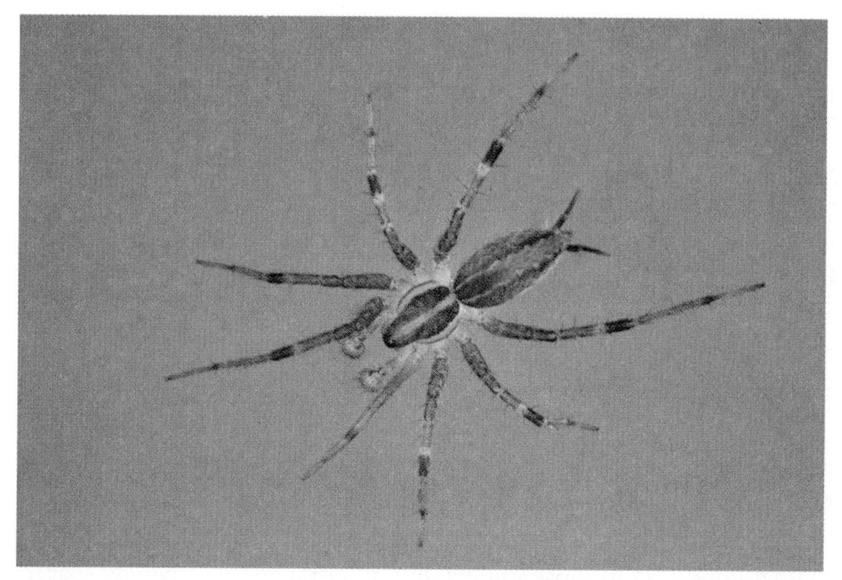

FIGURE 150. A male *Agelenopsis pennsylvanica* displays his enlarged palpal tarsi.

vegetation of forests, and lower stalks and limbs of bushes and weeds. In the center of the sheet web is a funnel retreat where the spider hides. Since the web contains no sticky threads, the spider depends upon his swiftness. When an insect falls onto the sheet web, the spider emerges instantly from its retreat, seizes the prey, and returns immediately to its retreat with a fresh meal. The spider lives for only one year, usually succumbing to the first severe winter freeze. Egg sacs are deposited on nearby vegetation. Spiderlings emerge in the spring and construct a new generation of sheet webs.This species ranges from New England south to Tennessee and west to Oregon and Washington (Kaston, 1978).

Size: Length of female 10 to 17 mm; of male 9 to 13 mm (Kaston, 1978).

FAMILY HAHNIIDAE—Hahniid Spiders

Hahniidae is a small family of three genera and 19 species in the U.S. The spiders of Hahniidae were originally classified with other sheet web weavers in the family Agelenidae (Comstock, 1912) but have since been segregrated as a separate family. Similar to the agelenid spiders, the hahniid spiders spin sheet webs; however, they never incorporate funnels or hidden retreats in their webs. The webs are only inches in diameter and are so delicate that they are difficult to detect, as they seem to lie on top of the ground between the stalks of thin grass or moss. Usually the spider can be collected from beneath its web.

The hahniid spiders are small spiders, some only 1.5 mm in length and the largest only 3.6 mm. The peculiar arrangement of spinnerets has warranted the placement of the hahniid spiders into their own family. The three pairs of dainty spinnerets are not clustered together in a longitudinal bundle as in most spiders, but are arranged in a widely spaced transverse row. A single broad spiracle is located well in advance of the spinnerets (Roth, 1993; Gertsch, 1934; Opell and Beatty, 1976). We represent this small family with a single species, *Neoantistea agilis*.

Family Hahniidae—*Neoantistea agilis* (Keyserling)

Identifying Characteristics: At first glance these spiders resemble the clubionids. However, the tarsi have three claws and lack claw tufts. More distinctive among the hahniid spiders are three pairs of transversely arranged spinnerets. The posterior pair of spinnerets is the most laterally placed and is the longest. It can be seen extending beyond the end of the abdomen in the photographed specimen. The abdomen of *N. agilis* is dark gray. A yellowish band with serrated edges extends down the dorsal midline of the abdomen. On the posterior one-third of the abdomen, the median band diffuses to form a series of four yellowish chevrons. The cephalothorax is light brown and in some individuals it has a reddish tint. The femora of each leg has the same background color of the

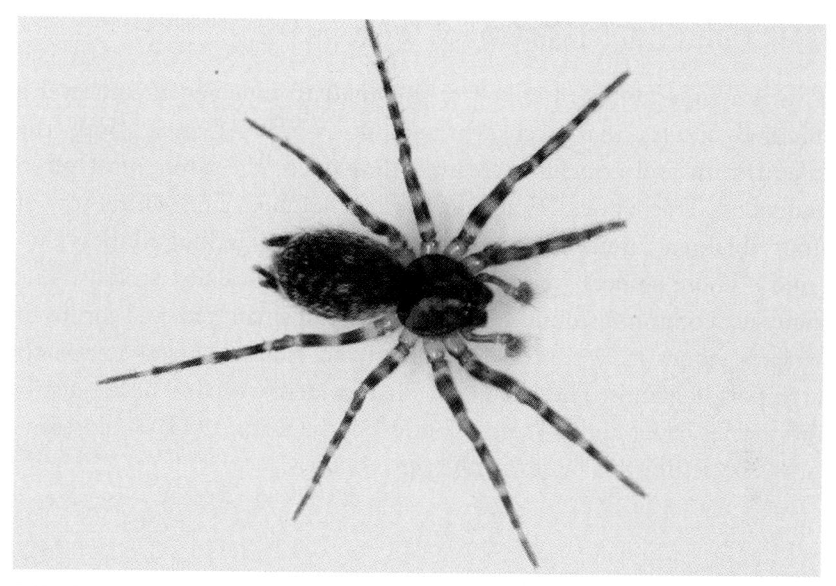

FIGURE 151. This *Neoantistea agilis* was collected beneath its tiny web on a dew-covered grass lawn. It was brought immediately into the laboratory and photographed against a white background.

abdomen. The distal ends of the legs, from patella to tarsus, are yellowish and are ringed with the brown colors of the cephalothorax.

Ecology and Behavior. This spider is most likely found in its tiny, delicate sheet webs which are spun near the ground and supported by grasses and moss. The webs are so delicate that they are seen best in grasses and cultivated fields when covered with morning dew. This spider is especially sensitive to water and during a drought its webs become a scarcity, and the spiders seek moisture and seclusion beneath shaded rocks. This spider is found throughout the U.S. and Canada (Kaston, 1978).

Size: Length of female 2.5 to 3.2 mm; of male 2.25 to 2.6 mm (Kaston, 1978).

FAMILY DICTYNIDAE—Meshweb Weavers

This is a large family of cribellate spiders with nine genera and over a hundred species that occur throughout North America. Both the calamistrum and cribellum are well-developed, the latter most often undivided. Trochanters III and IV are not notched. The ventral row of stout tibial macrosetae found in other cribellates is lacking in these dictynids. These spiders weave webs with zig-zag, hackled strands. The snares are commonly found among the twigs of small trees and shrubs or beneath stones in the forest floor. Dictynid spiders possess extremely large poison glands. These are very small spiders with 140 of the nearly 160 species being under 4 mm in body length (Roth, 1993). A representative species for this family is *Dictyna annulipes.*

FIGURE 152. Female *Dictyna annulipes* sits in her web constructed in the top of a dead twig.

Family Dictynidae—*Dictyna annulipes* (Blackwall)

Identifying Characteristics: This tiny spider will require the aid of a magnifying lens to see it well. This species has brownish-black legs, a brownish-black carapace which is covered with grayish-white hairs, and a black and white abdomen. The abdominal pattern of this wide-ranging species is highly variable. Kaston (1978) pictured two color-pattern variants of this species. Our specimen (see above) was photographed in northeastern Vermont. Its black abdomen is covered anteriorly and laterally by white hairs. Posteriorly and laterally, the white hairs become sparse and form three or four white lines which cross the abdomen to form a pattern of chevrons. The mid-dorsum of the abdomen is largely black and forms a broad stripe which courses the length of the abdomen, being crossed posteriorly by the white chevron lines. Near the middle of the abdomen, the black stripe forms two black spots, which extend laterally and superficially resemble "eyes". Perhaps the false eyespots are used

FIGURE 153. The small irregular "meshweb" of *Dictyna annulipes* in the terminal portion of a dead, defoliated weed.

to startle would-be predators. The male resembles the female but is more slender and has longer legs (Gertsch, 1979).

Ecology and Behavior: The *Dictyna* snare often consists of irregular cribellate webbing that entangles the terminal end of a twig or weed. However, the spider seems content to spin its web in and around tiny crevices of boards of fences and buildings. The above photograph depicts a *Dictyna* web anchored to the terminal branches of a dead, defoliated weed. The tiny dictynid spins its unique web by first laying down foundation lines from stem to stem. Then it weaves a criss-cross of viscid bands over this foundation. Thick, hackled webbing is characteristic of these cribellate spiders. Mating usually takes place during early summer. The lens-shaped egg sac is laid in late summer and the young hatch and spend some time in the web with mother before leaving during early fall (Gertsch, 1979). *Dictyna annulipes* ranges over most of the U.S., and most of Canada to Alaska (Kaston, 1978).

Size: Length of female 2.9 to 4.4 mm; of male 2.4 to 3.8 mm (Kaston, 1978).

FAMILY AMAUROBIIDAE—Hackledmesh Weavers

This family is characterized by having a bipartite cribellum. The calamistrum is usually less than one-half the length of the metatarsus. They are distinguished from the Dictynidae by their larger size, which is greater than 4 mm in body length. They can be distinguished from other stout-legged cribellates by the absence of trochanteral notches. The eight eyes are in two wide, nearly straight rows. There are 8 genera and 83 species in the U.S. The family is widely distributed across all of the U.S. (Roth, 1993). Chamberlin (1944) gave a summary of the known North American Amaurobiidae.

Family Amaurobiidae—*Callobius bennetti* (Blackwall)

Identifying Characteristics: This spider has a cribellum that is divided into two parts. A calamistrum is present in females and is about one-half the length of metatarsus IV. The calamistrum may be poorly developed and inconspicuous in males. The homogenous eyes are in two widely spaced and straight rows. The chelicerae are large and powerful and each is provided with a boss and scopula. Both anterior and posterior fang furrows bear teeth. The chelicerae and cephalic portion of the cephalothorax are black. The remainder of the cephalothorax, as well as the palps and legs, are golden brown without conspicuous spots or annuli. The abdomen is darker, being grayish-brown, with light yellow spots and about four chevronlike markings near the posterior end.

Ecology and Behavior: This spider prefers to live in cool, damp and dark situations. According to Comstock (1940), they have been found in

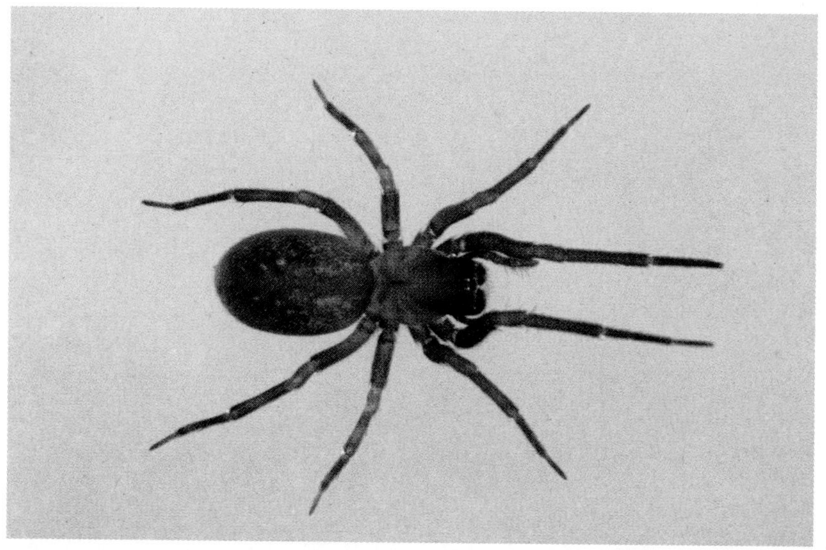

FIGURE 154. *Callobius bennetti* captured beneath a rotten log within its hackledmesh web retreat.

crevices in rock cliffs, in cellar walls, in hollow logs and under stones. Most of their webs are loose and irregular and seem to follow no particular construction plan; however, when their web is built around a crevice in a wall, the web lines seem to radiate outward from the retreat and support a sheet of hackled bands. According to Comstock (1940), the flattened egg sac contains a covering of loose webbing. It is attached to a stone or other object and covered with irregular silken threads. Egg sacs may be found during July and August. Kaston (1948) found two sacs during August, one containing 84 eggs and the other 144 eggs.

Size: Length of female 5 to 12 mm; of male 5 to 9 mm (Kaston, 1978).

Family Amaurobiidae—*Coras medicinalis* (Hentz)

Identifying Characteristics: The ground color of the entire body is yellow to amber. Upon this lighter background are distinctive patterns of dark gray stripes, spots, chevrons and annuli. On the cephalothorax is a

FIGURE 155. A female *Coras medicinalis.*

V-shaped mark extending from the eyes to the mid-region. Within the sides of the carapace is a second but broader V-shaped mark that extends to the pedicel. On the anteriormost one-fourth of the abdominal dorsum is a broad, dark median stripe. Posterior to this is a series of four or five dark chevron markings. The legs are marked with dark rings. Ventrally, the sternum has a distinct yellow median stripe bordered by dark gray. The chelicerae are large and robust, with a boss and scopula.

Ecology and Behavior: *C. medicinalis* builds an extensive, broad, platform-like web that is equipped with a tubular or funnel retreat. Their webs are often constructed beneath overhanging rock ledges where the platform of the web sags somewhat. The webs also may be found in and around hollow tree stumps, loose tree bark, under stones, and in various kinds of crevices. They are not uncommonly collected in basements and cellars. Adults and juveniles have been found hibernating in silken retreats under rocks and other objects (Moulder, 1992). *C. medicinalis* is found throughout the entire eastern U.S. It ranges from the eastern U.S. and adjacent Canada west to Minnesota and Texas (Kaston, 1978).

Size: Length of female 9.5 to 13.5 mm; of male 9 to 13 mm (Kaston, 1978).

FAMILY ANYPHAENIDAE—Ghost Spiders

The anyphaenid spiders are medium to small spiders that are white to dull gray in color. The pale whitish colors give them the common name "ghost spiders." They are often found as they hunt their prey among woodland foliage. They may also be found wrapped within a folded leaf held tightly with silken threads. The anyphaenids have claw tufts composed of flattened spatula-shaped hairs. Additionally, the tracheal spiracles on the abdominal venter are distantly removed from the spinnerets, being nearly midway the length of the abdomen.

The spiders of Anyphaenidae are similar in appearance and behavior to the spiders of the family Clubionidae. Authors of the past have grouped these two families together. However, arachnologists currently separate these families based upon the location of the tracheal spiracles and in the particular forms of the claw tufts. The family contains five genera and 37 species in North America (Roth, 1993). The three most common genera of the eastern U.S. are represented in this chapter. More specific information on this interesting family may be obtained by reading reports by Platnick (1974) and Dondale and Redner (1982).

Family Anyphaenidae—*Anyphaena celer* (Hentz)

Identifying Characteristics: This is a pale yellow or white spider in which the cephalothorax is darker than the abdomen or legs. On the cephalothorax two gray, broken longitudinal lines extend posteriorly from the posterior lateral eyes. On the abdomen are two faint longitudinal series of gray spots that run parallel to the midline and are separated by transverse rows of spots forming chevron-like markings. Taxonomically important is the spiracular furrow that is midway between the epigastric

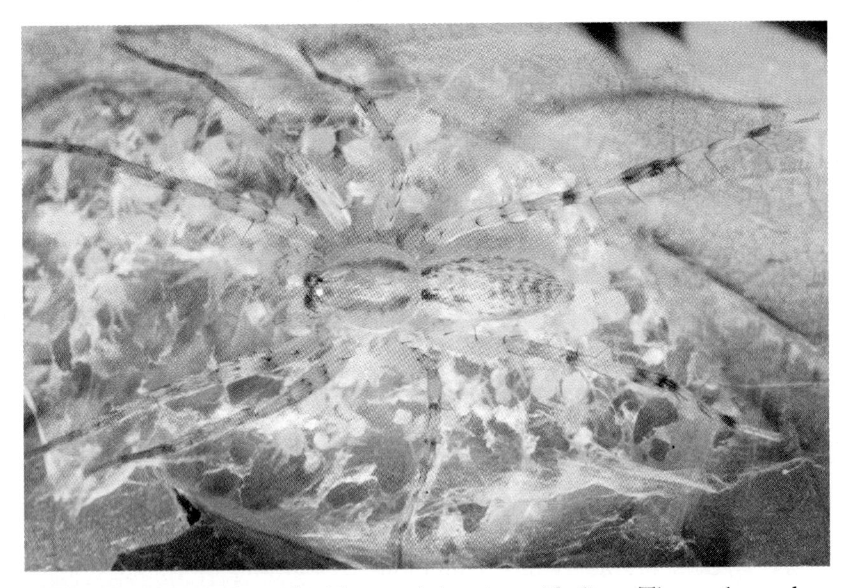

FIGURE 156. An anyphaenid spider guards her tiny spiderlings. The mother and offspring were found within a silken retreat within a folded leaf.

furrow and the spinnerets. Additionally, the anterior eyes are smaller than the posterior eyes. Segments of each leg are all the same color, yellow. From the underside of the male, coxae III and IV have many short bristles that look like stiff brushes.

Ecology and Behavior: This species may be found in the leaf litter of the forest floor, in grasses and bushes and under logs and rocks. Perhaps the easiest method to find specimens is to search for rolled-up leaves in low shrubs and bushes. The leaves may then be plucked from the vegetation and opened while being held over a net. During the opening of the leaf, the spider will often run out of the leaf with amazing speed, but hopefully it will fall into the collector's net. This anyphaenid is found from New England and adjacent Canada south to Alabama and west to Texas and Wisconsin (Kaston, 1978).

Size: Length of female 5 to 6 mm; of male 4 to 5 mm (Kaston, 1978).

FIGURE 157. Male *Anyphaena celer*.

Family Anyphaenidae—*Hibana gracilis* (Hentz)

Identifying Characteristics: This spider is yellow to white. It is distinguished from other members of the family by the brown chelicerae that are especially prominent against the pale-colored body. On the carapace is a pair of grayish longitudinal stripes. The abdomen is covered with red to black spots, arranged in two longitudinal rows. Scattered spots decorate the lateral regions of the dorsal abdomen. On the ventral surface, the spiracular furrow is much nearer to the epigastric furrow than to the spinnerets.

Ecology and Behavior: *Hibana gracilis* behaves much like the clubionid spiders. Both can be found living in silken nests in rolled-up leaves of herbs, shrubs and grasses. They may be collected from leaf litter in the forest and by sweep netting grasses and small bushes. Many specimens have been collected from houses and barns. The distribution of *H. gracilis* ranges from New England to Florida and west to Texas and Kansas (Kaston, 1978).

FIGURE 158. A male *Hibana gracilis* found in a folded leaf.

Size: Length of female 6.4 to 7.0 mm; of male 5.7 to 6.5 mm (Kaston, 1978).

Family Anyphaenidae—*Leupettiana mordax* (O.P. Cambridge)

Identifying Characteristics: This spider has long been known as *Teudis mordax* (Kaston, 1978). At first glance, this spider appears small and ant-like. Closer examination shows that the males have very large chelicerae that project forward a considerable distance. Additionally, the chelicerae are brownish to orange in color and armed with elongate fangs. The carapace is uniformly light red-brown and lacks the longitudinal bands typically seen on many other anyphaenids. The abdomen is lighter in color than the carapace and has transverse rows of dark brown spots, giving a banded appearance. The femora are darker than the other leg segments. Like other members of this family, *Leupettiana* has the spirac-

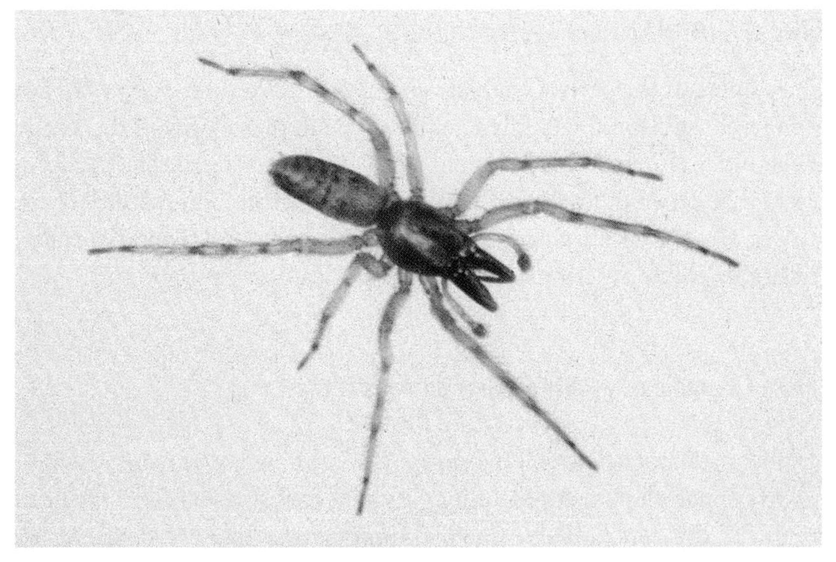

FIGURE 159. A male *Leupettiana mordax* showing the distinctively enlarged chelicerae. Also seen are the enlarged palpal tarsi.

ular furrow far forward from the spinnerets and they have the tip of the tarsus with claw tufts consisting of lamelliform (flattened) hairs.

Ecology and Behavior: We captured our specimens among tall grasses and foliage of low trees and underbrush in moist swampy lowlands near a large spring impounded to form a two-acre lake. These spiders were obviously hunting prey among shoreline grasses. We saw our specimens as they traversed silken threads connecting one tall grass patch to another. The spiders were running along these threads. When we tried to capture them, they quickly scurried away and made evasive movements much like those of an ant trying to avoid capture. *L. mordax* is distributed from Maryland south to Florida and west to southern California (Platnick, 1974).

Size: Length of female 3.9 to 5.5 mm; of male 3.7 to 5 mm (Kaston, 1978).

FAMILY LIOCRANIDAE—Liocranid Spiders

The spiders of the genera *Phrurotimpus, Phrurolithus* and *Scotinella* have been removed from the family Clubionidae and placed within the family Liocranidae (Roth, 1993; Jackman, 1997). Jackman stated that the characters that separate these two families "still seem to be unclear." These spiders, like clubionids, have eight eyes in two rows and two tarsal claws. Additionally, the anterior spinnerets are conical and contiguous.

Family Liocranidae—*Phrurotimpus borealis* (Emerton)

Identifying Characteristics: This spider is recognized by a shiny, orange-yellow cephalothorax. A pair of longitudinal dark lines outlines the head region of the cephalothorax, while another pair of lines delineate its curved margins. The ground color of the abdomen ranges from gray to purple. Due to a covering of iridescent scales, the abdomen glistens from

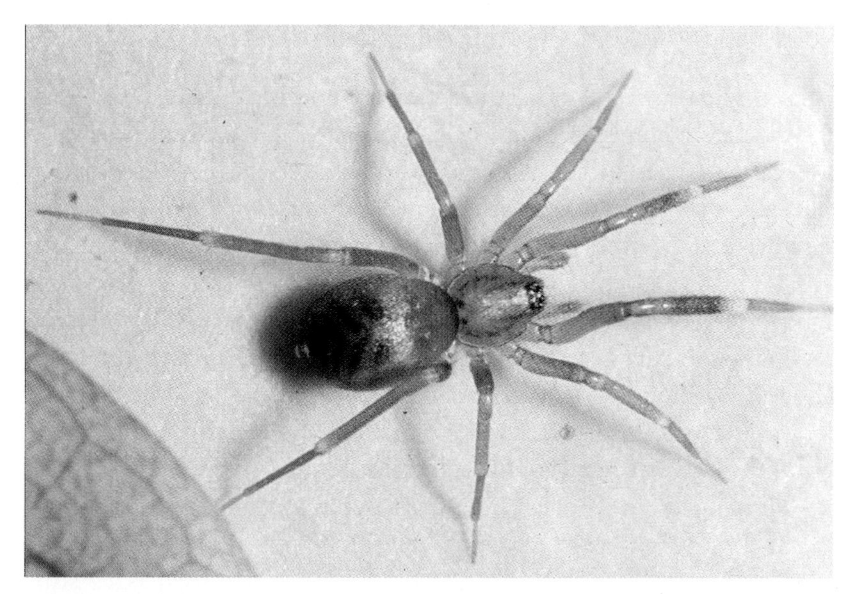

FIGURE 160. A female *P. borealis* which has lighter legs and a darker body than typical specimens.

green to pink, changing as the spider moves. From the mid-region to the spinnerets, the abdomen has four dark chevrons. The long, slender legs are lighter in color than the body. Leg I, however, has darkened tibia and patella and are armored with several spines on its ventral surface (Headstrom, 1973).

Ecology and Behavior: *P. borealis* lives in retreats constructed under stones or in rolled-up leaves in open areas and short grasses. When disturbed they are exceedingly quick, escaping to adjacent vegetation. When at rest, legs I and II are directed forward and bent so that patella and tibia overlie the cephalothorax; legs III and IV are directed to the rear. Bright red to brown circular egg sacs are secured to the underside of stones (Emerton, 1961). This spider is found east of the Mississippi River from southern Canada and New England south to Alabama.

Size: Length of female 2.4 to 3.6 mm; of male 2.8 mm (Kaston, 1978).

FAMILY CLUBIONIDAE—Clubionid Spiders

This is a large family of two clawed noctural hunting spiders. During the day they are commonly found in the foliage of small trees and shrubs or beneath stones, where they make tubular retreats. In the evening, they leave their retreats in search of prey in the litter of the forest floor or the foliage of the trees and shrubs. The family is separated from related families by having cylindrically shaped anterior spinnerets that are contiguous with, or directly in front of, the posterior spinnerets. The eyes are in two rows and, with few exceptions, homogeneous in shape and color. The legs are prograde, such that legs I and II are not held out from the body as in the crab spiders. Most species have drab, evenly colored bodies, while some have remarkably colorful markings on the abdomen.

This family has undergone considerable taxonomic restructuring over the last twenty years and many species do not fit nicely into either this family or the Gnaphosidae. Claw tufts were earlier used as a distinguishing feature of this family, but this character has since been shown to be quite variable. Currently, Clubionidae contains over twenty genera and 190 species. Clubionids are distributed commonly throughout the U.S. and Canada (Roth, 1993).

Family Clubionidae—*Cheiracanthium inclusum* (Hentz)—Agrarian Sac Spider

Identifying Characteristics: This spider is relatively pale in color with the cephalothorax and abdomen varying from pale beige to yellow and often with a tinge of green. A distinctive and darker colored lance-shaped cardiac mark is seen on the dorsal midline of the anterior abdomen. According to Kaston (1978), the body color varies according to the food eaten. The chelicerae are large, elongate and powerful, dark brown and conspicuous against the pale colored body. As with other members of the genus, the spider lacks a thoracic groove.

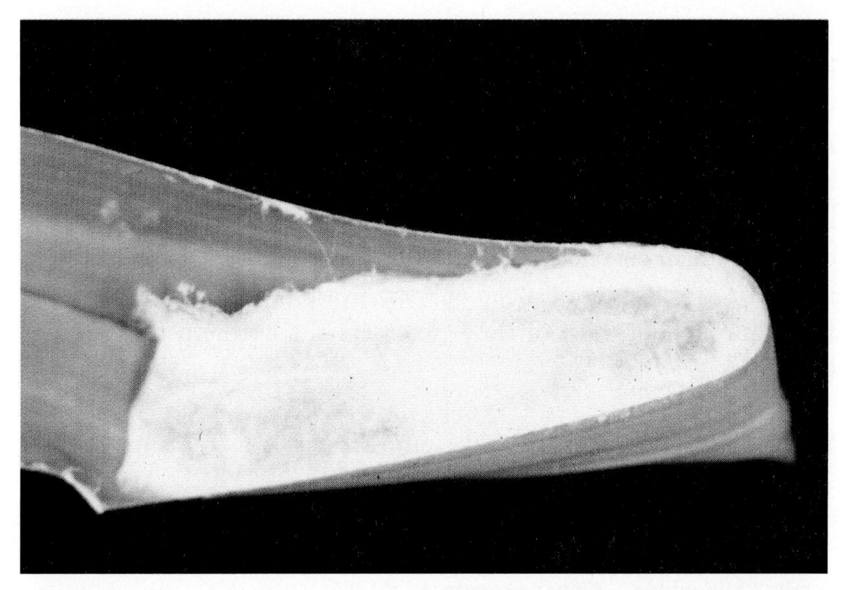

FIGURE 161. The tubular, silken retreat of a clubionid, *Cheiracanthium inclusum.*

FIGURE 162. *Cheiracanthium inclusum* has large, strong chelicerae and has been reported to envenomate gardeners who accidentally disturb them.

Ecology and Behavior: *C. inclusum* is commonly found running about on low trees and shrubs where they make silken tubular retreats in rolled up leaves. During the day the spider remains hidden in its retreat. At night, the spider emerges to become an active hunter. During the winter, their tubular retreats are constructed under stones and beneath the bark of trees. Supposedly, this spider has been responsible for several cases of human envenomation, but the bite is reported to be no worse than that of the sting of a bee or wasp (Kaston, 1978). This spider is found throughout most of the U.S. except the northern tier of states (Jackman, 1999).

Size: Length of female 4.9 to 9.7 mm; of male 4 to 7.7 mm (Kaston, 1978).

Family Clubionidae—*Elaver excepta* (L. Koch)

Identifying Characteristics: *E. excepta* is a creamy white ground spider that is distinguished from other clubionid spiders by its abdominal markings and legs. The creamy white abdomen is covered dorsally with long hairs and has a medial and two lateral rows of gray spots. Another pair of gray spots is on the underside of the abdomen, immediately anterior to the spinnerets. On the front legs the femur bears two distal prolateral spines. The legs are darker than the body except for the tibia, which is the same color as the abdomen. The cephalothorax is yellow in contrast to the whitish abdomen and legs.

Ecology and Behavior: *E. excepta* is most commonly encountered in forests beneath the cover of leaf litter, stones and loose tree bark. It is a very quick spider which ventures from its seclusion to hunt prey. It ranges from New England south to Georgia and west to Kansas and Nebraska (Kaston, 1978).

Size: Length of female 6 to 7.4 mm; of male 4.5 to 6.5 mm (Kaston, 1978).

FIGURE 163. A young male *Elaver excepta* pauses briefly in the forest litter.

FAMILY CORINNIDAE—Antmimic Spiders

Several spiders that were previously considered by Kaston (1978) and others to be in the family Clubionidae have been transferred into the family Corinnidae (Jackman, 1999; Roth, 1993; Reiskind, 1969). These spiders are ant mimics in both their physical appearance and in the manner in which they walk. They are nocturnal hunters and do not build snares. During the daytime, they may be found either beneath rocks, leaf litter, or in rolled-up leaves. Common genera now placed in this family include *Castianeira, Corinna* and *Trachelas.*

Family Corinnidae—*Castianeira amoena* (C.L. Koch)

Identifying Characteristics: From above, the ground color of the carapace and abdomen is orange. Additionally, the abdomen bears characteristic white and black markings that distinguish it from other *Castianeira* species.

FIGURE 164. An Antmimic spider pauses momentarily on a woodland leaf before scurrying away.

On the orange abdomen are nine black transverse bands; the anteriormost appears as a pair of connected triangular spots. Posteriorly, the black bands are often broken along the midline and those on the posterior half of the abdomen may be faint or lacking. The anterior half of the abdominal dorsum is scattered with white hairs. The legs are provided with alternating dark and light bands.

Ecology and Behavior: This clubionid is illustrated in other field guides as an "ant mimic" (Milne and Milne, 1980). In truth, it is easily mistaken for an ant based upon its running motions. Different from ants, it is always solitary and much faster in its escape. While standing, it often raises its first pair of legs, giving the impression of antennae and perfecting its mimicry. This spider is found in the evening and morning running upon the ground, venturing out to hunt small prey. When discovered it quickly seeks the cover of litter on the forest floor. *C. amoena* is found

FIGURE 165. A female *Castianeira amoena* pauses for a second while scurrying across a concrete sidewalk.

from North Carolina south to Florida and as far west as Texas and Kansas (Kaston, 1978).

Size: Length of female 7 to 8.8 mm; of male 5.7 to 6.8 mm (Kaston, 1978).

Family Corinnidae—*Castianeira descripta* (Hentz)—Red-spotted Antmimic

Identifying Characteristics: Spiders of this genus look and behave much like large ants. They are brown or black and the abdomen may be adorned with either white, red or orange colors. *C. descripta* has a longitudinal red band along the abdominal midline that gives it the appearance of a red mutillid wasp. The red band in some individuals is reduced to a series of red dots extending the length of the abdomen or limited to small red spots at the posterior end of the abdomen. The remainder of

FIGURE 166. The black body and broad red abdominal band distinguish *Castianeira descripta.*

the abdomen and cephalothorax is brown or black. The femora of the legs are brown or black, similar to the color of the body. The distal segments of legs I and II are yellow. Tibia I and II have two pairs of ventral spines. The fang furrow of the chelicerae has two teeth on the anterior margin and two on the posterior margin. The endites are flattened and lack depressions commonly seen among the antmimics of the genus *Micaria.*

Ecology and Behavior: This antimimic has been known to associate with ants, simulating them in behavior and interacting with ant workers. When at rest, they can be found beneath stones, leaves and logs. When disturbed they can run quite rapidly. They are frequently encountered in our homes, especially during the fall months as they seek out mates. We have collected several individuals from offices of our university campus. Eggs are deposited in flattened, disk-shaped cocoons that are enveloped with tough, parchment like silk. These cocoons have a metallic luster and are attached to the underside of stones. *C. descripta* is found in most every state east of the Rocky Mountains, and its range extends into southern Canada (Kaston, 1978).

Size: Length of female 8 to 10 mm; of male 6.2 to 7.6 mm (Kaston, 1978).

Family Corinnidae—*Castianeira longipalpus* (Hentz)

Identifying Characteristics: This distinctively marked spider and other members of its genus were, until recently, placed within the family Clubionidae (Roth, 1993; Reiskind, 1969). For the most part, *C. longipalpus* is dull in color but with very contrasting and distinctive markings. The distal three-fourths of legs I and II are light colored while the femora are black. Leg IV has alternating light and dark rings. The cephalothorax is

FIGURE 167. *Castianeira longipalpus* is distinguished by the unique patterns of light and dark markings on each pair of its legs.

chestnut brown. The abdominal dorsum is black and crossed by seven or fewer transverse white bands. In some individuals there may be as few as only one abdominal white band, and this is near the posterior end. The thoracic groove is well marked.

Ecology and Behavior: This species mimics carpenter ants in appearance and behavior. It has even been found living in association with these ants. In mimicking ants, they move about slowly with the same head and abdominal gestures as the ants. It moves its anterior legs to simulate the antennae movements characteristic of ants. *C. longipalpus* does not make a snare, but rather constructs a retreat in rolled-up leaves, under rocks, or in litter. We have collected most of our specimens beneath rocks in pastures and power-line easements. Eggs are laid in disk shaped cocoons that are secured to the underside of stones. It is a nocturnal species. This spider is found from New England and adjacent Canada south to Florida and west to Oklahoma and North Dakota and on into the Pacific Northwest (Kaston, 1978).

Size: Length of female 7 to 10 mm; of male 5.5 to 6.1 mm (Kaston, 1978).

Family Corinnidae—*Trachelas tranquillus* (Hentz)

Identifying Characteristics: *T. tranquillus* is a common spider that is likely to wander into your home or office during the summer months. This species is identified by its reddish brown carapace and sternum which are thickly covered with minute punctures. The abdomen is pale white to yellow and bears a gray cardiac mark on the dorsum. The anteriormost legs take on the reddish brown color of the cephalothorax while the more posterior legs are colored similarly to the abdomen. Kaston (1978) reports that this species is unique in the genus in that the anterior eyes are separated by a diameter of the height of the clypeus.

FIGURE 168. Female *Trachelas tranquillus* exhibiting its red brown carapace with small punctures and a pale white abdomen. This spider has been suspected of human envenomation (Uetz, 1973).

Ecology and Behavior: *T. tranquillus* is commonly collected in forest leaf litter during the summer and autumn months. In the southern states, it often finds its way into homes during June and July. It is most active in the evening as it stalks prey. It runs quickly and seeks refuge in the camouflage of the brown leaf litter. Specimens can be collected during the day under the loose bark of trees and from folded leaves. *T. tranquillus* ranges throughout the eastern United States from New England and adjacent Canada southwest to Georgia and west to Minnesota and Kansas (Kaston, 1978).

Size: Length of female 6.8 to 10 mm; of male 5 to 6.1 mm (Kaston, 1978).

FAMILY GNAPHOSIDAE—Ground Spiders

The members of this family have their eyes in two rows. The posterior median eyes are flattened and oval or irregularly shaped. The chelicerae have a boss. The tracheal spiracle is located close to the spinnerets. The anterior spinnerets are cylindrical, almost parallel, widely separated and longer than the posterior spinnerets. The legs usually bear spines and the tarsi have a brush of hairs on the lower surface referred to as a scopula. The tarsi bear two claws and claw tufts. Most species have drab, homogeneously colored bodies, while some have remarkably colorful patterns.

Gnaphosids usually construct a silken, tubular retreat in a rolled-up leaf or under stones. Most species stalk and hunt prey at night and seek the seclusion of their retreats during the day.

This family contains over twenty genera and nearly 250 species (Roth, 1993). Gnaphosids are distributed commonly throughout the U.S. A taxonomic treatise on the family Gnaphosidae was done by Platnick and Dondale (1992).

FIGURE 169. The Parson Spider, *Herpyllus ecclesiasticus,* is a ground dwelling spider that often enters garages, houses and buildings.

Family Gnaphosidae—*Callilepis pluto* (Banks)

Identifying Characteristics: Members of this genus are recognized by the bright orange-brown cephalothorax and similarly colored legs. The abdomen is bluish black with contrasting white markings. Three broad white bands cross the abdomen. Upon close examination, the posterior lateral eyes are clearly larger than the posterior median eyes. Furthermore, the posterior row of eyes is only slightly recurved and about the same length as the anterior eye row. The chelicera has a narrow lamina on its retromargin and lacks serrated teeth as occurs among the closely-related spiders of the genus *Gnaphosa.*

Ecology and Behavior: *Callilepis* are very active small spiders found in pastures and forest litter. They are usually found under stones and when discovered they quickly elude capture by disappearing into the surrounding litter. Both sexes spin small irregular webs beneath stones where females

FIGURE 170. A male *Callilepis pluto* revealing its distinctive abdominal markings.

deposit a round, convex egg sac. *Callilepis* ranges from New England south to Alabama and west into the states of the Rocky Mountains and the Pacific Northwest (Kaston, 1978).

Size: Length of female 4.8 to 6.2 mm; of male 3.8 to 5 mm (Kaston, 1978).

Family Gnaphosidae—*Cesonia bilineata* (Hentz)

Identifying Characteristics: This is a distinctive gnaphosid having black and white dorsal stripes that extend the length of the thorax and abdomen. This spider has a broad median white stripe bordered by two dorsolateral black stripes, extending from the eyes to the spinnerets. White stripes extend along lateral aspects of the thorax and abdomen. The ventral side of the abdomen is dirty white and has two lateral black stripes. The white regions of the body are covered with silvery white

FIGURE 171. Female *Cesonia bilineata* displaying its contrasting black and white stripes.

hairs. The cephalothorax is much narrower anteriorly than it is in the thoracic region. The anterior and posterior rows of eyes are nearly straight. The legs are gray. Like most gnaphosids, the anterior spinnerets are cylindrical, longer than the posterior spinnerets and widely separated. A taxonomic revision of the genus *Cesonia* was done by Platnick and Sadab (1980).

Ecology and Behavior: *C. bilineata* can be found under rocks and leaf litter in wooded areas. When uncovered, it quickly scurries away and can change directions instantly without altering its pace. This makes this spider difficult to catch. In spite of its elusiveness, this spider is curious and shows little fear of humans when captured. *C. bilineata* often wanders into houses and may be seen crawling on the walls and ceilings and may even come down to investigate the human inhabitants below. Drawn by its curiosity, this spider may even crawl over one's body. *C. bilineata* is common from New England to Georgia and west to Nebraska (Kaston, 1978).

Size: Length of female 6 to 8 mm; of male 5 to 6 mm (Kaston, 1978).

Family Gnaphosidae—*Drassyllus depressus* (Emerton)

Identifying Characteristics: The cephalothorax is reddish-orange and this strongly contrasts with the dark grayish to black abdomen. Unlike most spiders, the dark abdomen is without any discernible dorsal markings. The legs are yellowish-brown. The posterior eye row is procurved with the posterior median eyes large, oval and obliquely oriented, and are much closer to each other than to the posterior lateral eyes. The promargin of the cheliceral fang furrow has three to six teeth and the retromargin has two to four (Kaston, 1948). The genus is widespread geographically with most species occurring in the western states. According to Platnick and Sadab (1982) the genus has 44 species and many are difficult to separate and identify without tedious dissection of

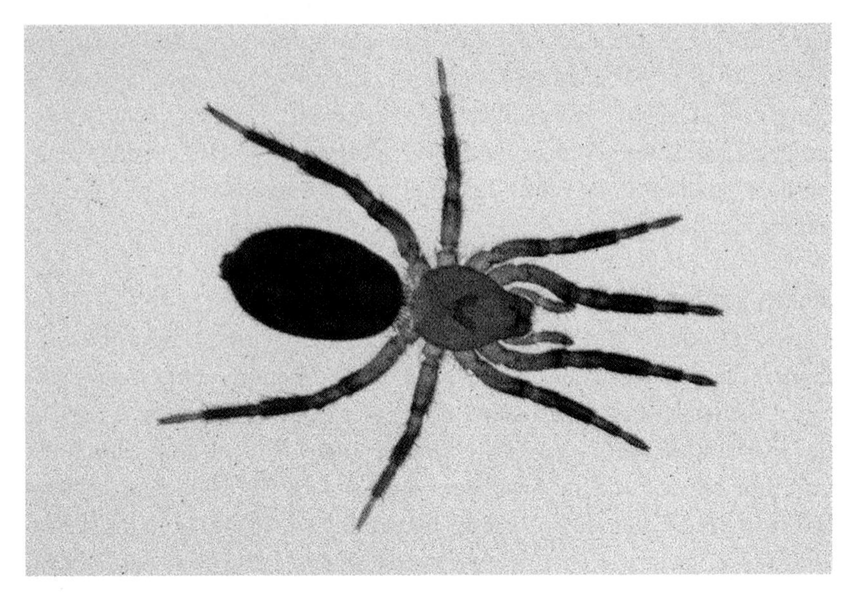

FIGURE 172. Female *Drassyllus depressus.*

the genitalia. A taxonomic revision of the genus *Drassyllus* was done by Platnick and Sadab (1982).

Ecology and Behavior: These spiders live on the ground under leaves, stones and logs. We have collected this species in the same type of habitat as *Herpyllus ecclesiasticus, Cesonia bilineata* and *Sergiolus capulata,* i.e., very hot and dry clay hillsides beneath stones and leaves. One method that we have used to collect this spider is with a common garden hose equipped with a small nozzle to create a hard jet of water. The powerful stream of water is used to systematically sweep the habitat and drives the spiders from beneath their hideouts so that they scurry ahead of the water where they may be collected into plastic bags. They are very fast and make quick ant-like movements. These spiders usually hunt their prey along the ground during the night and usually remain hidden during the daytime. *D. depressus* is found from New England and adjacent Canada southward to Georgia and westward to Arizona, Colorado and Oregon (Kaston 1978).

Size: Length of female 4.7 to 6 mm; of male 5 mm (Kaston, 1978).

Family Gnaphosidae—*Herpyllus ecclesiasticus* (Hentz)—Parson Spider

Identifying Characteristic: This tiny spider has a very distinctive color pattern. It body is largely black with white markings extending down the midline of the back. The cephalothorax is black along the sides and has a wide, dull white stripe down the midline. The abdomen is similarly colored being black along the sides, but with a much whiter midline stripe.

FIGURE 173. The Parson Spider displaying its distinguishing abdominal markings.

The median white stripe is constricted about midway of the dorsum to form an upside-down V. There, the white stripe is interrupted by black, just before it terminates anterior to the spinnerets. The ventral side of the abdomen is black along the sides and white in the midline. A taxonomic revision of the genus *Herpyllus* was done by Platnick and Sadab (1977).

Ecology and Behavior: The Parson Spider is a ground spider, being found most often in wooded areas under rocks, old boards and debris. It hibernates in the space beneath loose tree bark. In late fall it deposits a flat, white egg sac in its retreat, attached to the tree trunk. When disturbed, it runs exceedingly fast making astonishing maneuvers in order to avoid predators. It is a difficult spider to catch because of its incessant movements. *H. ecclesiasticus* ranges from New England to Georgia and west to Oklahoma and Colorado (Kaston, 1978).

Size: Length of female 8 to 13 mm; of male 5.5 to 6.5 mm (Kaston, 1978).

Family Gnaphosidae—*Micaria aurata* (Hentz)

Identifying Characteristics: This gnaphosid is yellowish-orange to brown all over, except for legs I and II which are light yellow distally from the femur. The elongate cephalothorax is smooth and uniform in color with only the slightest thoracic groove, if not lacking altogether. Similar to *Micaria longipes,* the orange-brown abdomen is covered with iridescent scales so that it gives a red or green metallic sheen in bright light. The abdomen bears two pairs of white transverse bands, one at the base of the abdomen and the other at mid-length. The spider resembles an ant, and in some specimens this resemblance is enhanced by the abdomen, which may be slightly constricted in the mid-region. The posterior abdomen may possess five or six dark chevrons, which Emerton (1961) attributed to the southern variety.

FIGURE 174. *Micaria aurata* revealing its distinctive abdominal iridescence and chevrons.

Ecology and Behavior: This is an exceptionally active spider that prefers dry, sandy habitats. They may be collected from dunes of beaches to grassy plots where they hunt prey during the hottest part of the day. It is more common in southern states, but extends into New England and west to Texas and Nebraska (Kaston, 1978).

Size: Length of female 5.2 to 6.7 mm; of male 5 to 6 mm (Kaston, 1978).

Family Gnaphosidae—*Micaria longipes* (Emerton)

Identifying Characteristics: This spider looks and acts like an ant. The general color is yellow-brown to reddish-orange and is covered with iridescent scales. The abdomen has two pairs of thin transverse spots, one pair anterior and the second, more conspicuous, in the mid region. The

FIGURE 175. Female *Micaria longipes* exhibiting its distinctive abdomen, which has the anterior one-half reddish-orange while the posterior one-half is black.

posterior third of the abdomen is the darkest part of the spider, being dark brown or black. In the past, spiders of the genus *Micaria* have been grouped in the family Clubionidae because the spinnerets are contiguous, or the anterior spinnerets are aligned directly in front of the posterior spinnerets. It differs from clubionids and has been regrouped with the gnaphosids because the posterior median eyes of *Micaria* are oval and different in size from the other eyes. Similar to other gnaphosids, *Micaria longipes* has an oblique depression on the ventral surface of the endites. A taxonomic revision of the genus *Micaria* was done by Platnick and Sadab (1988).

Ecology and Behavior: These spiders run like ants over the ground in dry areas. Because of this behavior and their nesting habits, they are often confused with spiders of the genus *Castianeira*. *Micaria longipes* has been reported to have a northern distribution from the northern states west to

Utah (Kaston, 1978). The photographed specimen was collected in northcentral Alabama.

Size: Length of female 5 to 6 mm; of male 5 mm (Kaston, 1978).

Family Gnaphosidae—*Sergiolus capulatus* (Walckenaer)

Identifying Characteristics: The carapace is bright orange becoming a dark, burnt orange color around the eyes. The abdomen is black with three transverse white bands crossing the dorsum. The middle white band has a T-shaped mark extending forward from its center. The legs are mostly orange except for the following: femora I and II are black, and black rings are seen at the distal end of femur IV and both ends of tibia IV. The sternum is light orange. The venter of the abdomen is black with a broad, median light stripe that is pointed behind. This

FIGURE 176. The tiny ant-like spider, *Sergiolus capulatus.*

species was previously placed in the genus *Poecilochroa* (Jackman, 1997; Platnick and Sadab, 1981).

Ecology and Behavior: This colorful little spider superficially resembles an ant. The bright orange colors, coupled with its ant-like movements, might lead one to mistakenly identify it as a small mutilid wasp. We have found this spider most often in dry areas beneath small rocks, leaf litter and woodland debris. Upon uncovering its hiding places, the tiny spider moves with exceeding rapidity across the ground and often escapes the grasp of the collector. We have also found that this spider often lives in groups with other members of its own species as well as other look-alike spider species within the same genus (e.g. *Sergiolus ocellata*). We have collected this spider using a garden water hose. Using a powerful stream of water, the ground is sprayed in order to dislodge and overturn stones, leaves, and debris. This uncovers these spiders which run about to escape the water. They are then captured with ease. They are especially common during the summer and early fall under warm weather conditions. *S. capulata* is found from New England south to Georgia and west to Oklahoma and Nebraska (Kaston, 1978).

Size: Length of female 10 mm; of male 5.5 to 7.0 mm (Kaston, 1978).

FAMILY SPARASSIDAE—Giant Crab Spiders

The sparassid crab spiders resemble the thomisid crab spiders in their behavior, by having a dorsoventrally flattened body, and by having a lateral positioning of their legs. However, they differ by having two straight rows of eyes and with powerful chelicerae armed on the lower margin with large teeth. These giant crab spiders are mostly tropical or subtropical and are found in the southern part of the U.S. These spiders are nocturnal predators. **Beware, some biologists have claimed that their bite is venomous.** These spiders have been placed in the family Heteropodidae by some authors (e.g., Jackman, 1997, 1999).

Family Sparassidae—*Heteropoda venatoria* (Linnaeus)—Huntsman Spider

Identifying Characteristics: This large spider is yellowish brown to tan in color with a transverse band of white hairs extending across the front of the face. A similar transverse band is seen along the hind margins of the cephalothorax. The anterior median eyes are smaller than the anterior laterals. Tibia I has three or four pairs of ventral spines.

Ecology and Behavior: The huntsman spider is found worldwide in tropical regions. This tropical species apparently has been introduced into the United States and has taken up residence in the warmer southern states from Florida to Texas (Kaston, 1978). It is supposed that the huntsman was accidentally introduced into the U.S. in bunches of bananas from

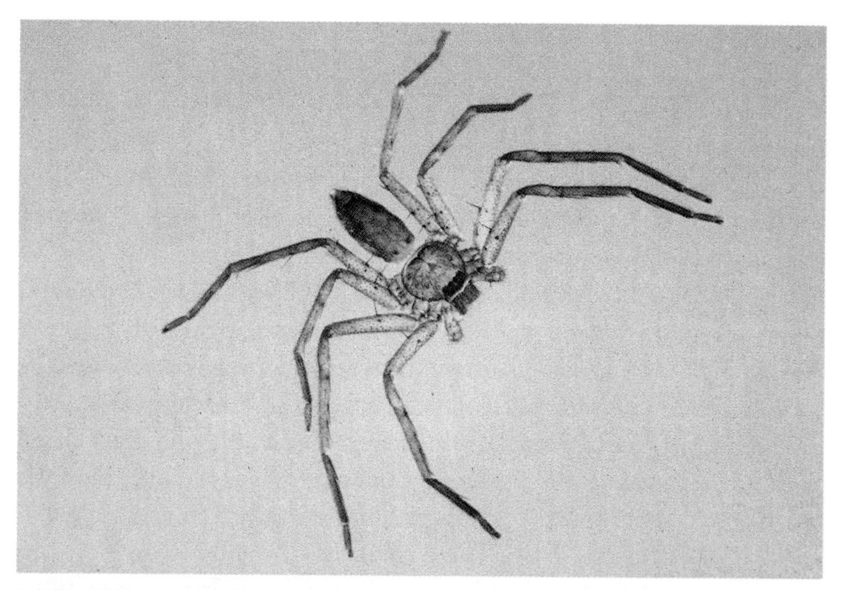

FIGURE 177. *Heteropoda venatoria,* a species of giant crab spider which has been accidentally introduced into the southern U.S. via shipments of goods from Central America.

native habitats in Central America. We have found numerous specimens in a receiving and distribution warehouse for tropical aquarium fishes and aquatic plants in north Alabama. Undoubtedly, these specimens found a free ride to the U.S. in the shipping containers.

Huntsman spiders are very swift, aggressive and elusive. They can move with unbelievable speed and disappear into the narrowest cracks and crevices. This spider is so fast that it appears to vanish from sight the moment that the collector's hand reaches out to grab it. According to Comstock (1942), the primary food of this large spider is cockroaches. The female lays her eggs in a flat, pinkish, cushion-like egg sac that is three-fourths inch wide and one-fourth inch thick. The female carries the egg sac beneath her body. The young spiderlings emerge through a slit in the margins of the egg sac.

Size: Length of female 23 mm; of male 20 mm (Kaston, 1978).

FAMILY PHILODROMIDAE—Running Crab Spiders

Spiders of this family are crab-like by having a dorsoventrally-flattened body and by holding the first and second pairs of legs out from the body. Similar to the thomisid crab spiders, the philodromid spiders are very active and aggressively pursue their prey. Most philodromid species live on vegetation and seek hiding places beneath either stones or tree bark. When at rest the body is held close to its substrate. It usually selects a background that matches its often grayish-to-mottled body coloration, thereby rendering it superbly camouflaged against its background.

The philodromids are distinguished from the closely related thomisid crab spiders in that their second pair of legs is longer than the first. The third and fourth legs are as long or nearly as long as the first pair of legs (Emerton, 1912). The tarsi of the first and second legs have ventral hairs that are broad and feather-like and each leg has claw tufts. Evidence on chromosome number, internal eye anatomy and development of the spiderlings warrant the separation of philodromids and thomisids into two distinct families. Philodromidae contains five genera

FIGURE 178. Philodromid spiders are distinguished by Leg II being longer than I, III, or IV.

that range throughout much of North America (Roth, 1993). Philodromids have been well described in a taxonomic treatise by Dondale and Redner (1978).

Family Philodromidae—*Philodromus vulgaris* (Hentz)

Identifying Characteristics: The general body coloration of this philodromid is gray and brown with dark spots, bands, stripes and blotches. The cephalothorax is slightly wider than long. A Y-shaped mark is evident on the carapace. The abdomen is grayish brown and covered with short, brown hairs. Anteriorly, it has a lanceolate median stripe; posteriorly, it has a herringbone pattern. From a lateral view, the abdomen is higher anteriorly and posteriorly and depressed midway. From above, the abdomen widens at three-fourths its length and then tapers posteriorly

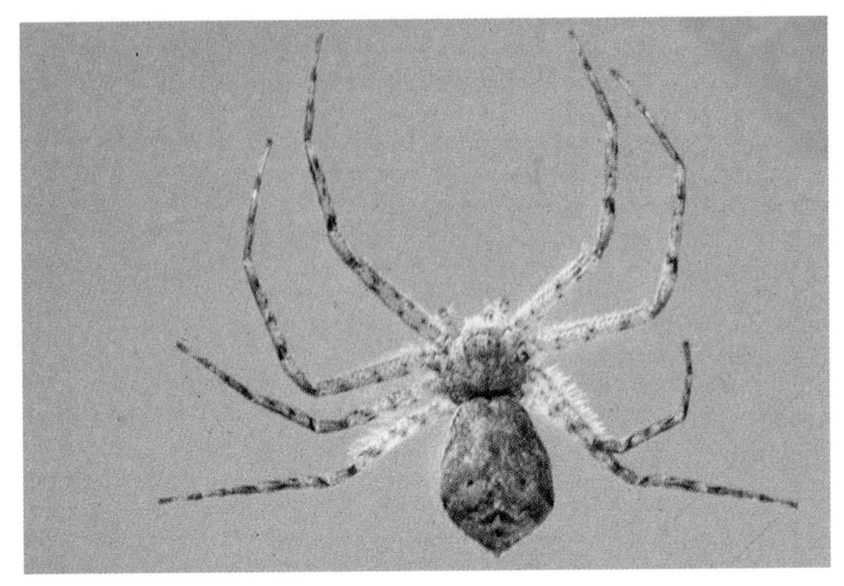

FIGURE 179. A female of *Philodromus vulgaris.* Note leg pair II which is slightly longer than leg pair I, a characteristic that distinguishes the running crab spiders from the thomisid crab spiders.

to a blunt point. The legs are tan with dark brown bands which darken near each joint.

Ecology and Behavior: This cryptically colored, flattened spider is admirably adapted for life on tree trunks, wooden fences and boards of unpainted houses. Its grayish-brown mottled body colors blend it into its surroundings so well that it is nearly impossible to see. In addition to its protective coloration, the spider is superbly adapted for squeezing into crevices and skittering around vertical tree trunks with amazing agility. The spider is fast and difficult to collect. Sexual maturity is attained during April and most adults survive through October. A female may produce up to five egg sacs per season; the earlier sacs contain forty to fifty eggs, but the later sacs as few as seven eggs (Kaston, 1948). The egg sacs are attached to stones, twigs or leaves and are guarded by the female.

P. vulgaris ranges throughout the eastern U.S., extending from New England south to Florida and west to the Rocky Mountains (Kaston, 1978).

Size: Length of female 4.5 to 8 mm; of male 4.5 to 6.3 mm (Kaston, 1978).

Family Philodromidae—*Thanatus vulgaris* (Simon)

Identifying Characteristics: The genus *Thanatus* differs from the closely-related genus *Philodromus* by having the legs almost all the same length. When measured, legs II and IV are slightly longer than legs I and III. Of taxonomic importance, female *T. vulgaris* have three spines on the lateral surface of femur I. In males, there is only one dorsal spine on the tibia of the palps. The cephalothorax has a light beige to brown middorsal, dagger-shaped stripe, bordered on each side by a lighter, pale-yellowish stripe about the same width as the midline stripe. This is bordered on

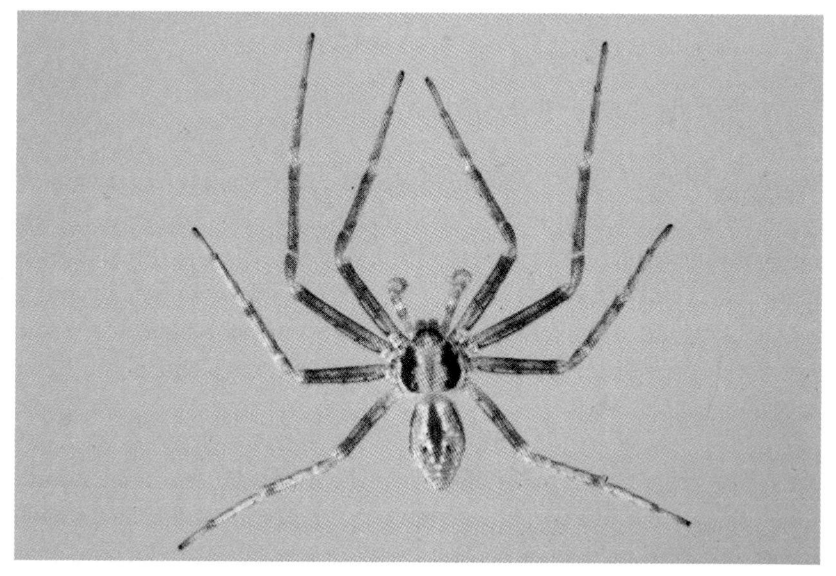

FIGURE 180. A male *Thanatus vulgaris*.

each side by a contrasting dark, slightly broader stripe. Lateral to this and forming a border to the cephalothorax is a thin pale-yellowish stripe. This spider has a distinctive dark lanceolate mark in the cardiac region of the abdomen. This dorsal abdominal mark is bordered by a thin, white line forming a distinctive V. Three pairs of dark pinpoint dots are present on the abdominal dorsum beside the lanceolate mark.

Ecology and Behavior: These spiders are common in prairie-like situations, either in tall grass or broomsedge pastures. The photographed specimen is unusual in that it was collected from Spanish moss from live oak trees of coastal Alabama. It was so well-camouflaged in the moss that it was discovered only by vigorously shaking a handful of moss into a white cloth insect net. These spiders will sit motionless on vegetation in easy view, differing from the clandestine behavior of true crab spiders. When provoked, they move with exceptional briskness. They will capture and feed on smaller spiders and insects. *T. vulgaris* is found from Ohio south to Georgia and west to Idaho and California (Kaston, 1978).

Size: Length of female 5 to 10 mm; of male 4.5 to 6 mm (Kaston, 1978).

Family Philodromidae—*Tibellus duttoni* (Hentz)

Identifying Characteristics: *Tibellus duttoni* is a dirty yellow to gray philodromid spider. A dark brown band extends the length of both the cephalothorax and abdomen. Along the posterior half of the abdomen the median band becomes less distinct and a pair of black spots straddles the median band. The elongate abdomen is five times longer than wide. In a closely related species, *T. oblongus,* the abdomen is only about three times longer than wide.

Ecology and Behavior: Spiders of the genus *Tibellus* are commonly found in tall grasses and bushes. They seem to prefer broomsedge and straw-like grasses that are beige in color. When at rest, they sit elongate on a stem or leaf with legs I and II stretched forward and legs III and IV

FIGURE 181. A female *Tibellus duttoni* rests on a piece of wood while awaiting its prey.

pointed backward. Their position affords them excellent camouflage when they are in their "sit and wait" foraging strategy. When startled, they use their long legs to vault from plant to plant, eventually ducking out of sight. Because of their superb camouflage, they are best collected with a sweep net. They are found from New England south to Florida and west to Texas and Minnesota (Kaston, 1978).

Size: Length of female 6 to 10 mm; of male 5 to 7 mm (Kaston, 1978).

FAMILY THOMISIDAE—Crab Spiders

Crab spiders are medium to small spiders that have a horizontally flattened body. They are typified in that the first two pairs of legs are much enlarged, rotated laterally, and powerful as compared to the third and fourth pairs of legs. These spiders are not only crab-like in appearance but also in behavior. They can run rapidly with considerable maneuverability in the forward, backward, and lateral directions. They walk and run primarily on the third and fourth pair of legs, and they use their enlarged first pair for seizing prey.

The anterior lateral eyes and posterior lateral eyes of the thomisid crab spiders are closely placed and are mounted upon elevations. In some species the lateral eyes are on a conjoined tubercle (Kaston, 1978).

Crab spiders do not build webs, snares, retreats, or molting nests. They typically wander among vegetation and forest litter, hunting by stealth and ambush. They select their vegetation for the best possible camouflage and are often difficult to see unless they should move. Crab spiders are best collected by sweeping tall grass with insect nets or by vigorously shaking foliage into a white net or cloth sheet.

Thomisidae is a large family that contains nine genera and over a hundred different species common to the United States (Roth, 1993).

FIGURE 182. Female White Banded Crab Spider, *Misumenoides formosipes,* sits on the yellow petals of a garden flower awaiting its prey.

Those readers desiring more detailed information on these fascinating spiders should consult the works of Gertsch (1939; 1953), Schick (1965) and Dondale and Redner (1978).

Family Thomisidae—*Bassaniana versicolor* (Keyserling)—Bark Crab Spider

Identifying Characteristics: This brown crab spider is frequently confused with those of the genus *Xysticus*. However, *Bassasiana versicolor* is distinguished by the presence of a prominent unpigmented declivity on the cervical groove from which lighter streaks radiate toward the margins of the carapace. From an anterior view, the anterior row of eyes is marginally recurved and the crest of the carapace is relatively flat. The abdomen is brown or gray with darker margins. Along its length the abdomen bears alternating light and dark markings. In dark individuals the dark spots are so large that the entire spider appears nearly black. The legs of lighter individuals are covered with dark spots which may congregate at the end of the joints.

Ecology and Behavior: This is a common species found on fences and wooden bridges. Its brown color blends well with these unpainted substrates, making it difficult to spot. The spider often seeks seclusion beneath loose bark on trees and underneath stones. *Bassaniana versicolor* occurs throughout the eastern half of the United States, from Canada to Florida and west to Arizona.

Size: Length of female 5.5 mm; of male 4.5 mm (Kaston, 1978).

Family Thomisidae—*Misumena vatia* (Clerck)—Goldenrod Crab Spider

Identifying Characteristics: The carapace of *Misumena vatia* is white to yellowish brown but the ocular region is chalky white and often tinged with red. The lateral margins of the carapace are usually slightly darker

FIGURE 183 A-B. *Bassaniana versicolor* showing its (**A**) recurved anterior eye row, (**B**) carapace declivity and abdominal markings.

FIGURE 184 A-B. Female (**A**) and (**B**) male *Misumena vatia.*

than the median region. The abdomen is the same color as the carapace and often lacks markings. When abdominal markings are present, there are anteriolateral red bands. The legs are lighter in color than the body. The male goldenrod spider has a red to green-brown carapace that often bears a lighter medial band extending to the eyes. The abdomen has broad red or red-brown bands. In the male, legs I and II are red-brown and legs III and IV are spotless white or yellow. *Misumena vatia* is often confused with *Misumenoides formosipes,* however, the former lacks the prominent ridges in the clypeal region.

Ecology and Behavior: This is one of the most abundant flower spiders in the United States, second only to *M. formosipes. M. vatia* is most likely to be on white or yellow flowers, such as goldenrod. As it moves to a slightly different colored flower, it is capable of changing its basal color. This camouflaged spider awaits its prey, usually flower pollinating insects, in the flower heads and secures them by ambush. This crab spider preys upon a wide variety of insects including moths, flies, bees and hemipterans, which are often larger than the spider. *M. vatia* is found throughout the United States and Canada (Kaston, 1978).

Size: Length of female 6 to 9 mm; of male 2.9 to 4 mm (Kaston, 1978).

Family Thomisidae—*Misumenoides formosipes* (Walckenaer)—White Banded Crab Spider

Identifying Characteristics: *M. formosipes* varies in overall body color from yellow to yellowish-brown to creamy white. On top of its basic yellow color, the lateral margins of the carapace are darker and medially they often bear bright red spots. Red spots also occur on the dorsal surface of legs I and II. The abdomen is sometimes unmarked; however, red or brown markings are more common. The most diagnostic feature of this spider are the prominent white ridges above and below the front row of eyes. Additionally, there are few spines over the legs and body and the anterior and posterior lateral eyes are distinctly separated and are not confluent.

FIGURE 185. Female of *M. formosipes.*

Ecology and Behavior: *M. formosipes* is the most common of the flower spiders. Flower heads such as goldenrods, asters, sunflowers, and fleabanes are its favorite lairs. Pasture grasses and prairie provide it with its best habitat. In late summer, as hordes of flying insects converge upon the yellow and white flower heads, the whitebanded crab spider sits patiently, camouflaged amongst the blossoms, with its unsuspecting prey at easy reach. To some extent, they are capable of changing body color depending upon flower color. Adults mature in July and August and deposit up to 100 eggs in a white, silken egg sac among the vegetation in September. Young emerge from the egg sac in the early spring. *M. formosipes* is common throughout the entire continental United States and Canada (Kaston, 1978).

Size: Length of female 5 to 11.3 mm; of male 2.5 to 3.2 mm (Kaston, 1978).

Family Thomisidae—*Misumenops asperatus* (Hentz)—Northern Crab Spider

Identifying Characteristics: This species closely resembles in color, size and ecology the other flower crab spiders belonging to the genera of *Misumenoides* and *Misumena. M. asperatus* is distinguished from these other spiders in that the carapace, abdomen, and legs are distinctively covered with numerous short stiff spines. Additionally, the anterior eyes of *M. asperatus* differ in size such that the laterals are larger than the medials. The carapace is yellow and has broad brownish longitudinal lateral bands. The ocular region is white. On the anterior half, the abdomen has red streaks laterally. On the posterior half the abdomen has a mottled brown to red-brown V-shaped mark pointed toward the posterior. All legs are yellow, except that tibia and metatarsus I bear red annuli.

Ecology and Behavior: *M. asperatus* is frequently collected from grass and foliage by using a sweep net. While sitting in yellow to white blooms, their camouflage is superb and they are very difficult to see, even with close scrutiny. The Northern Crab Spider is widespread throughout most of North America (Kaston, 1978).

Size: Length of female is 4.4 to 6 mm; of male 3 to 4 mm (Kaston, 1978).

Family Thomisidae—*Synema parvula* (Hentz)

Identifying Characteristics: This tiny spider is handsomely colored. The thorax is yellowish-orange with the lateral margins being dark brown. The eyes are encircled by light rings. The abdomen is white or light yellow anteriorly but has a wide black to brown band traversing the posterior one-fourth. The black band does not quite reach the spinnerets. On the anterior part of the top of the abdomen are small dark spots and often some white markings. A dark band is present on each side of the venter of the abdomen. The legs are yellowish-orange with pairs III and

FIGURE 186 A-B. Female (**A**), having just captured a green leafhopper; and male (**B**) Northern Crab Spider, *Misumenops asperatus.*

FIGURE 187. Female of *Synema parvula.*

IV lighter in color than I and II. Legs III and IV are not more than two-thirds as long as legs I and II. The thorax is as wide as it is long and strongly convex. The abdomen is as wide as long, being widest across the middle, and somewhat pointed posteriorly. While the males are smaller than females, there are little differences in their color pattern.

Ecology and Behavior: We have collected this spider only by using a sweep net in tall grasses and the flower heads of herbaceous plants during the late summer and fall months. Hentz, who described this species, found it hiding among the blossoms of umbelliferous plants. This species is distributed from New Jersey and Illinois south to Florida and west to the Rockies (Kaston, 1978).

Size: Length of female 2 to 3 mm; of male 2.3 mm (Kaston, 1978).

Family Thomisidae—*Tmarus angulatus* (Walckenaer)

Identifying Characteristics: Like other crab spiders, legs I and II are enlarged and both are held forward in a lateriograde position. Legs I and II are darker and with more prominent annuli than leg III or IV. The clypeus (space between chelicera and eyes) is very conspicuous by being strongly sloped and nearly vertical. The abdomen is higher than the carapace and rises posteriorly to a tubercle. The size of the tubercle varies from individual to individual but is usually conspicuous when viewed from the side. *T. angulatus* is dark brown, mixed with yellow and gray. A median lighter band extends the length of the body from the clypeus to and over the posterior abdominal tubercle. The median band is crossed by red and dark brown chevrons or lines.

Ecology and Behavior: *T. angulatus* stalks its prey from plants, frequently camouflaging itself as it tightly clings to brown colored twigs. As this

FIGURE 188. Lateral view of *Tmarus angulatus* showing the pointed tubercle on its posterior abdomen.

spider clasps onto small twigs, its peculiarly pointed abdomen projects outward resembling a small bud. When prey comes within reach, it jumps with great swiftness and seizes the prey with its front legs. When disturbed, it can jump to the ground and run with great quickness. It occurs throughout the entire United States and southern Canada (Kaston, 1978).

Size: Length of female is 4.5 to 7 mm; of male 3 to 5 mm (Kaston, 1978).

Family Thomisidae—*Xysticus transversatus* (Walckenaer)—Tan Crab Spider

Identifying Characteristics: The carapace is relatively high. The anterior eye row is moderately recurved. The cervical groove is indistinct. The carapace of the female is tan to gray-brown with lighter areas medially.

FIGURE 189. The female Tan Crab Spider, *Xysticus transversatus*, guards her egg sac.

The abdomen is the same color as the carapace and has three dorsal pairs of dimples. The abdomen is flattened dorsoventrally and is broader than long. It is widest posteriorly and overhangs the carapace anteriorly. The abdomen has four transverse light band-like markings. The carapace and abdomen of the male are a darker brown than that of the female. The legs of males bear faint annulations. *Xysticus* characteristically bears many spines on the anterior carapace. Tibia I has three pairs of spines on its medial border. The anterior and posterior lateral eyes are on tubercles that are clearly separated.

Ecology and Behavior: *X. transversatus* is perhaps the most common cosmopolitan crab spider in the eastern U.S. It, however, is not restricted to a particular habitat, being found beneath rocks, under bark, or awaiting ambush of prey on a tree or in flowers and weeds. While hunting it often succeeds at capturing prey many times its size, such as butterflies and moths. *Xysticus* spiders are commonly collected in sweep nets. Otherwise, they are well camouflaged and difficult to detect. We have spotted this spider on vegetation as it carried and fed upon a moth three times its size. It occurs from New England south to Georgia and west to Texas and the Rocky Mountains (Kaston, 1978).

Size: Length of female 6 to 7 mm; of male 5 to 6 mm (Kaston, 1978).

Family Thomisidae—*Xysticus triguttatus* (Keyserling)—Three Banded Crab Spider

Identifying Characteristics: This moderately-sized *Xysticus* is yellowish brown with three pairs of spines on the prolateral surface of tibia I. The lateral eyes are non-confluent. It is distinguished within its genus by having a median light band extending the length of the carapace. The thoracic groove lies in this median band and bears a black spot. The lateral margins of the carapace are darker and sprinkled with black spots. Anteriorly, the eyes are situated on creamy white tubercles. The dirty white abdomen has a pair of small black spots anteriorly and three

FIGURE 190. Female Three Banded Crab Spider camouflaged on a dead limb.

transverse rows of black bands posteriorly. The posterior transverse bands may be broken along the midline. The male is usually darker than the female.

Ecology and Behavior: The Three Banded Crab Spider is commonly found among and within buildings and houses. However, its natural habitat is under the leaf litter of the forest floor. When inactive it seeks seclusion under tree bark or flat stones. Mating occurs in late spring following which the females deposit egg sacs in silken retreats. The retreats are usually made in folded leaves in which the female guards her egg sac until young emerge. Its distribution extends along the east coast of the United States, west to the Rocky Mountains, and north into Canada (Kaston, 1978).

Size: Length of female 4 to 6 mm; of male 3 to 5 mm (Kaston, 1978).

FAMILY SALTICIDAE—Jumping Spiders

The jumping spiders represent one of the most distinctive and easily recognized families of spiders. They are typically short, stout spiders with a broad cephalothorax. Most species are brightly colored or with distinct markings. The short and robust legs allow the spider to make exceedingly quick and agile forward, backward, or side-to-side movements. If provoked, they can jump distances several times their body length. Jumping spiders have the best vision among the spiders, second only to certain wolf spiders. Their eyes are arranged into three rows over the cephalon and in the shape of a quadrangle, called the ocular quadrangle. The four large eyes of the first row are anteriorly directed. The medial eyes in the first row are the largest of the eight eyes, such that they give the impression of automobile headlights. Eyes of the second row are very small and often difficult to see. Two large eyes of the third row demarcate the posterior limit of the ocular quadrangle. These eyes are directed upward and

FIGURE 191. Jumping spiders typically have short robust legs and enlarged front row eyes.

sometimes backward. Behavioral studies have shown that jumping spiders have good vision at distances up to several inches. They do have color vision and some depth perception, made possible by the simultaneous use of different eyes. Jumping spiders are aggressive diurnal predators. During the day, they may be found venturing from their den that they improvise from cracks and crevices in rocks and walls. Many species of jumping spiders may be found on the underside of vegetation, silently awaiting their prey. Salticids do not construct snares, although many species do make silken tubes folded within leaves. Their retreats are used for molting, overwintering, or depositing eggs. Jumping spiders often use silken drag-lines to regain their positions after making distant jumps.

A list of the jumping spiders of the U.S. and Canada was presented by Richman and Cutler (1978). A key to the genera was given by Roth (1993). Edwards and Hill (1978) gave useful photos of representative salticids.

Family Salticidae—*Anasaitis canosa* (Walckenaer)

Identifying Characteristics: The palps are black with iridescent white tibia and patella. Anteriorly, the cephalothorax is black and distinctly marked with white scales. Posterior to the third row of eyes, the scales form a pair of large white ovoid spots. Anterior to the third row of eyes, the scales form a pair of small white spots. Posteriorly, the cephalothorax is chestnut brown. Four large, dark, irregular spots on the abdominal dorsum can be used to recognize this spider. In some individuals, as in the photographed specimen, the two spots of each side may be fused. Additionally, a brown and white herringbone pattern extends from the black spots to the posterior end of the abdomen. In some specimens this herringbone pattern is daubed to a uniform gray-brown. Markings between the sexes are very similar; however, the abdomen of the male is narrower. Taxonomically important to this genus are the wide set eyes of the first row and more closely set rear eyes. Thus, the ocular quadrangle is a unique trapezoid shape. *Anasaitis canosa* has been called *Stoidis aurata* by Comstock (1912).

FIGURE 192. Female *Anasaitis canosa* displaying distinctively colored iridescent white palps.

Ecology and Behavior: *Anasaitis canosa* is commonly found in small groups or families from May to the first frost of winter and generally will reside under stones, logs or beneath loose tree bark. In an abandoned construction site, a total of six individual spiders of both sexes defended their territory when we disturbed their domain of old wood and culvert pipes. This species was initially described as a Floridian species (Comstock, 1912) but has since been found in other southeastern states, extending from South Carolina west to Texas (Kaston, 1978; Jackman, 1999).

Size: Length of female 5.3 to 5.8 mm; of male 5 to 5.2 mm (Kaston, 1978).

Family Salticidae—*Eris aurantia* (Lucas)

Identifying Characteristics: This moderately-sized jumping spider is hallmarked by its iridescent scales and white markings. The iridescent scales render many variations in color depending upon the direction of light. On the cephalothorax a characteristic band of white scales extends from the anterior lateral eyes around the carapace to the posterior declivity. The ocular quadrangle is black and the thorax is red-brown. A white scaly band outlines the base and lateral margins of the abdomen. Dorsally, the abdomen is orange-brown and has four pairs of black spots interrupted by three pair of white spots. The legs are a light chestnut color except Leg I, which has dark bands on the distal femur and tibia.

Ecology and Behavior: This spider lives among low shrubs and grasses where it hunts for a wide variety of insect prey. We have collected this spider mostly during late fall after the first short period of near freezing weather. During this time, the spider is often collected within a silken encasement wrapped up within a folded leaf. We have collected only a few specimens of this spider in the eastern U.S. and it does not appear to be a very common spider. It ranges from Delaware west to Illinois, south to Florida and west to Arizona (Kaston, 1978).

Size: Length of female 8 to 12 mm; of male 7 to 10 mm (Kaston, 1978).

Family Salticidae—*Eris marginata* (Walckenaer)

Identifying Characteristics: This species has been known previously as *Eris militaris* (Hentz), the Bronze Jumper. The males of this genus have large chelicerae extended in front of the body. In the male, leg I is comparatively large and thickly fringed with bristles on its medial surface. The male of this species has a black cephalothorax and a metallic bronze abdomen, thus its former common name. The cephalothorax and abdomen both have white bands on their lateral margins. A large white

FIGURE 193 A-B. (**A**) Female; and (**B**) male of *Eris aurantia*.

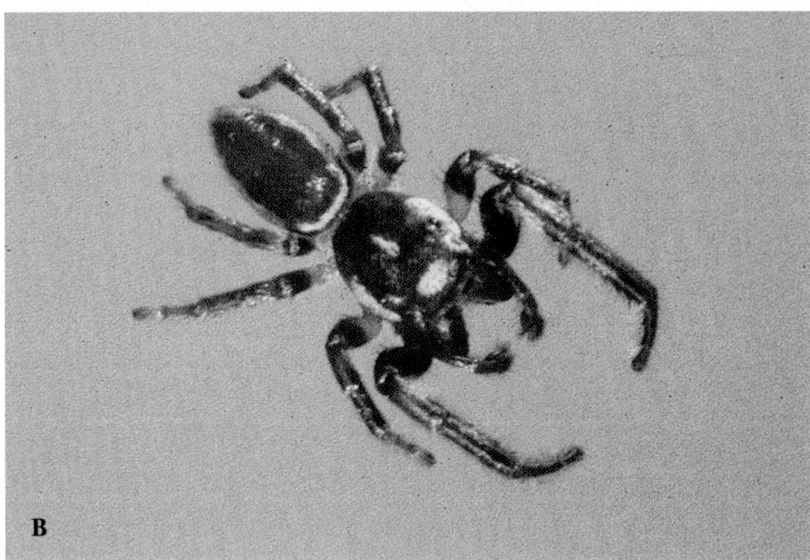

FIGURE 194 A-B. (**A**) Female; and (**B**) male of *Eris marginata*.

spot lies anterior to the third row of eyes. On legs II, III, and IV the coxa and the medial half of the femur are yellow. The female has a lighter colored cephalothorax that lacks the lateral white bands. The males of *E. marginata* are very similar in appearance to males of *Marpissa lineata.* Both species display lateral and central white markings on the cephalothorax and yellow femurs. However, the bronze colored abdomen and the presence of only two white abdominal bands distinguish this spider. The genus *Eris* has many taxonomic resemblances to *Phidippus* and *Metaphidippus.* The intermediate size and hairiness of *Eris* distinguish it from *Phidippus* (larger and hairier) and from *Metaphidippus* (smaller and less hairy).

Ecology and Behavior: *E. marginata* is frequently found on the limbs and leaves of shrubs and small trees. Its range extends throughout the United States and southern Canada. It, however, is one of the more frequently encountered *Eris* members in the eastern states, westward to Texas.

Size: Length of female is from 6 to 8 mm; the male is 4.7 mm.

Family Salticidae—*Habrocestum pulex* (Hentz)

Identifying Characteristics: *Habrocestum pulex* is a distinctive jumping spider in that it possesses an elongate cephalothorax, at least 1.5 times longer than wide. The abdomen is short and broad, often shorter and much broader than the cephalothorax. This species is best recognized by the white triangle on the thorax, with the base of the triangle bridging the posterior eyes and the apex of the triangle directed toward the pedicel. The abdominal markings are best described as irregular white spots. Individuals show considerable variation in their abdominal markings; however, there seem to be at least three distinctive consistencies. Anteriorly, the abdomen has white markings surrounding dark markings along the midline. Further posteriorly, there is a double white chevron. At the end of the abdomen is a series of small white and brown angular marks.

FIGURE 195. The tiny jumping spider, *Habrocestum pulex.*

Legs III and IV are nearly twice as long as legs I and II. All legs have alternating white and brown bands.

Ecology and Behavior: This small jumping spider (rarely more than 5 mm in length) exists in varying shades of gray and brown that give it excellent camouflage against the ground. When startled, this little spider will freeze and, when motionless, seeing it is difficult. *Habrocestum pulex* is commonly found leaping about wherever suitable camouflage is to be found, such as gray stones and dried leaf litter. Its range encompasses the entire eastern U.S. from New England and Canada to Florida and west to Louisiana and Nebraska.

Size: Length of female 4.5 to 5.5 mm; of male 4 to 4.5 mm (Kaston, 1978).

Family Salticidae—*Habronattus sp.*

Identifying Characteristics: Spiders of this genus are small and chunky. The cephalothorax is high and convex and about as long as wide. The legs are distinctive with the order from longest to shortest being 3, 1, 4, 2. Special ornamentations occur in the male *Habronattus* (photographed below) on Leg I and/or Leg III. These ornamentations are variously shaped, colored projections from the patella and fringes of hairs from the tibia. The male displays his ornaments before the female as a sexual lure during courtship. Markings on the cephalothorax and abdomen are usually more distinct in the male, consisting of longitudinal stripes and chevrons. The genus was recently revised by Griswold (1987).

Ecology and Behavior: This is a diverse genus with a total of 95 species, all of which are confined to Central and North America. The range of many species overlaps the eastern states. Most species are found in open woodlands in bushes and tall grass. The photographed male *Habronattus* was collected in October by sweeping tall road side grasses.

Size: Length of female *Habronattus* 5 to 7 mm; of male 4 mm to 6 mm (Kaston, 1978).

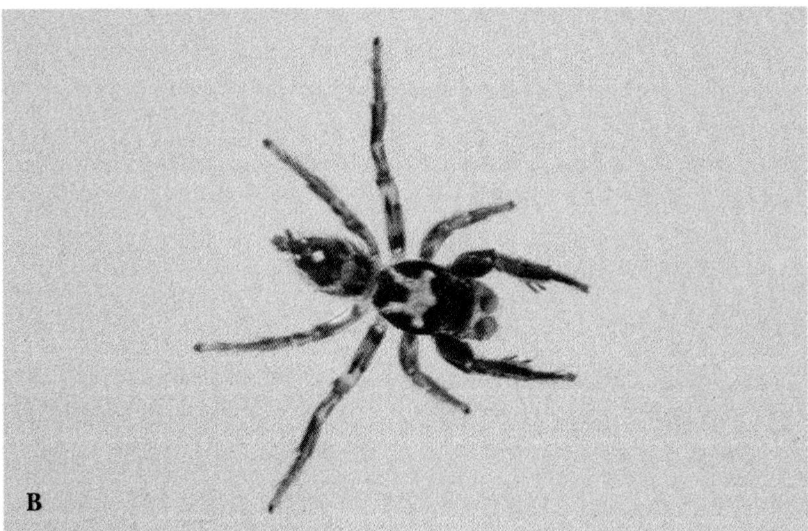

FIGURE 196 A-B. (**A**) Female *Habronattus agilis;* and (**B**) male *Habronattus* sp. with cephalothorax markings similar to the shape of a Japanese gazebo.

Family Salticidae—*Hentzia palmarum* (Hentz)—Longjawed Jumping Spider

Identifying Characteristic: The common name of this small spider is based upon the appearance of the male. The enlarged chelicerae extend from the body and are over half as long as the cephalothorax. In the male the first pairs of legs are enlarged, powerful, and dark brown while the other legs are much lighter in color. The third and fourth rows of eyes in the male are set upon dark spots, and white and yellow scales mark the clypeus. The female lacks the enlarged chelicera and the legs are translucent white to yellow. The abdomen of the female is often marked with brown chevrons and triangles. *H. palmarum* has been previously called *Attus ambigua* (Walckenaer) and *Wala palmarum* (Hentz) (Comstock, 1940). The genus *Hentzia* was revised by Richman (1989).

Ecology and Behavior: *Hentzia palmarum* is typically found on trees and shrubs of deciduous forests and is often collected from sweep netting of tall grasses in forest clearings and along roadsides. *H. palmarum* is distributed from New England and adjacent Canada southward to Florida and westward to Oklahoma and Nebraska (Kaston, 1978). The photographed female specimen was collected from a southern deciduous forest.

Size: Length of female 4.7 to 6 mm; of male 3.7 to 5.5 mm (Kaston, 1978).

FIGURE 197 A-B. (**A**) Female *Hentzia palmarum* with abdominal pattern of black triangles; (**B**) male, showing distinctive chelicerae.

Family Salticidae—*Lyssomanes viridis* (Walckenaer)—Magnolia Green Jumper

Identifying Characteristics: As both the common and species names infer, this spider is distinctively green. Indeed, it is bright translucent green in life (Roth, 1993). The eight eyes occur in four rows of two eyes each. Four black tubercles on the head bear the second, third and fourth rows of eyes. The arrangement of the eyes in this genus has suggested to many arachnologists that *Lyssomanes* is one of the most primitive of jumping spiders. This has led some arachnologists to place this spider into a separate family, the Lyssomanidae (Kaston, 1978). Extending the length of the abdomen are usually four pairs of dark spots. The chelicerae of the male are greatly enlarged and brown in color.

Ecology and Behavior: *Lyssomanes viridis* is a very active and fearless spider. Its green color provides excellent camouflage against the spring and summer growths of the vegetation of low bushes and trees. Neither sex shows fear of humans but has been described to dance for people as though it were at center stage for entertainment. The photographed specimen skillfully evaded and fought its collectors for several minutes from the camouflaging leaves of an elderberry bush. These green spiders can be spotted by their shadows if the collector stands beneath a tree or shrub and views through the fleshy, translucent leaves against the sun. Several species of *Lyssomanes* are common throughout Central America. However, only *L. viridis* is common to the southern United States. Its range extends from North Carolina south into Florida and west to Texas (Kaston, 1978).

Size: Length of female 7 to 8 mm; of male 5 to 6 mm (Kaston, 1978).

FIGURE 198 A-B. (**A**) Female; and (**B**) male Magnolia Green Jumper, *Lyssomanes viridis.*

Family Salticidae—*Maevia inclemens* (Walckenaer)—Dimorphic Jumper

Identifying Characteristics: The photographed specimen below typifies the mature female. The coxae of the legs are yellowish in color and translucent. The cephalon is dark brown and the eyes are black. On the thorax is a white mark somewhat in the shape of the Maltese cross of the German Catholic Church. The abdomen is dark brown and has several white transverse chevrons. The anteriormost abdominal chevron is broadest and more posterior chevrons are successively narrower. The male of the species can be of two distinct types. One is similar to the female while the other is quite different. A dark male variety is uniformly black (cephalothorax, palpi, and abdomen) and covered with shiny dark greenish scales. The legs are a transparent white. Distinctive in all males are three tufts of bristles on the cephalon between the first and third rows of eyes. Taxonomically, the genus *Maevia* is similar to *Marpissa* by having four pairs of ventral spines on Tibia I and by possessing teeth on the retromargin of the cheliceral fang furrow. Upon casual examination, *Maevia* is the much larger spider. This species is also known under the name *Maevia vittata* (Hentz).

Ecology and Behavior: We have collected most of our specimens from the sides of buildings and vegetation in residential areas. It also occurs in open woodlands. It can also be collected by sweep netting roadside grasses and weeds. It is a curious spider and shows a great deal of interest in its collector, moving from side to side and turning its head so as to get a better look at its captor. It is found from New England and adjacent Canada to Florida and westward to Texas and Wisconsin (Kaston, 1978).

Size: Length of female 6.5 to 10 mm; of male 4.8 to 7 mm (Kaston, 1978).

FIGURE 199 A-B. The Dimorphic Jumper, *Maevia inclemens*. (**A**) male; (**B**) female.

Family Salticidae—*Marpissa lineata* (C.L. Koch)—Fourlined Jumper

Identifying Characteristics: *Marpissa lineata* is typified among the jumping spiders by a wide, flat cephalothorax. The cephalon is dark brown and the eyes are widely placed, laterally. A central white spot sits between the posterior eyes. The thoracic region of the cephalothorax is red. The abdomen, characteristically, has four longitudinal white lines. One pair of the white lines is dorsal and visible from above. The other pair of white lines is ventrolateral and visible best from below. Males are easily recognized by their yellow legs, except tibia I that is contrastingly dark brown. In the female the legs are homogeneously brown. Of taxonomic importance, *Marpissa* has four pairs of ventral spines on tibia I. *Marpissa lineata* may be mistaken for a similarly colored *Phidippus* jumping spider; however, *Marpissa lineata* is one-third the size of most *Phidippus* species.

FIGURE 200. Male of *Marpissa lineata.*

Ecology and Behavior: The photographed specimen was collected during July as it stalked its prey from the foliage of an elderberry bush, growing wild along a lake shore. *Marpissa lineata* is common throughout the eastern states excluding Florida. Its western range extends to the Mississippi River; its northern range extends to New England and Canada (Kaston, 1978).

Size: Length of female 4 to 5.3 mm; of male 3 to 4 mm (Kaston, 1978).

Family Salticidae—*Marpissa pikei* (G & E Peckham)—Pike Slender Jumper

Identifying Characteristics: This spider is easily recognized from other jumping spiders by its elongate abdomen, which is more than three times as long as wide. The first pair of legs is enlarged and not used for walking but held out from the body. The cephalothorax is generally brownish in color. The abdomen in both sexes has a middorsal longitudinal dark brown to black stripe bordered on each side by a white band. *Marpissa pikei* was referred to as *Hyctia pikei* by Comstock (1912).

Ecology and Behavior: These peculiarly elongate jumping spiders reside in tall grasses where they sit motionless and well concealed among the plant stems. They are voracious spiders that will seize and kill other arthropods of equal size or smaller. We have collected and placed a living female in a jar with other small insects. She immediately began killing and feeding on these small arthropods. These spiders are most likely collected in sweep nettings through tall grasses. Even after being dumped into a pan from the net, they can be easily overlooked unless they move. When they do, they do so with exceedingly quick and jerky movements. Their range extends from New England and Canada south to Florida and west to Nebraska and Arizona (Kaston, 1978).

Size: Length of female 6 to 9.5 mm; of male 6 mm to 8.2 mm (Kaston, 1978).

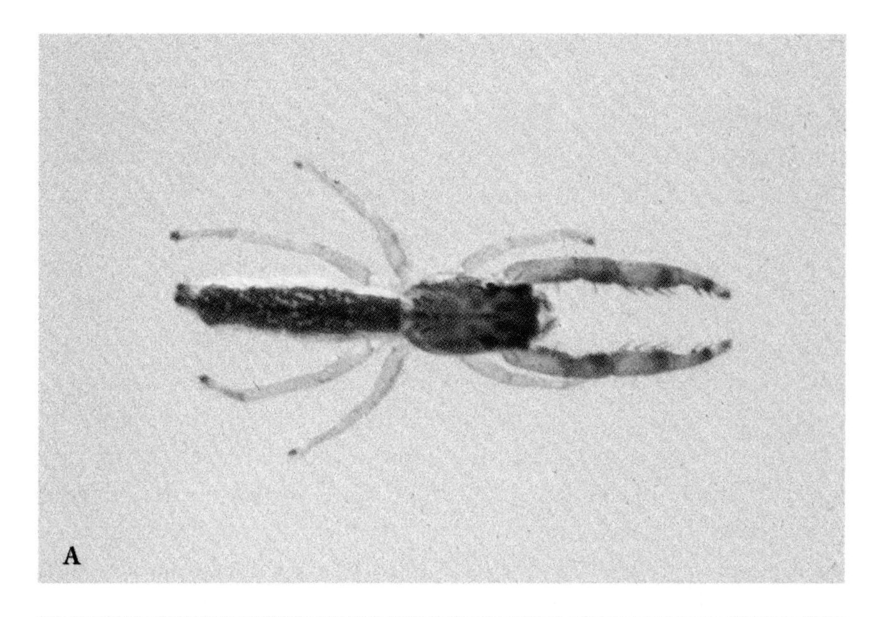

FIGURE 201 A-B. (**A**) male; and (**B**) female of *Marpissa pikei*, showing unique elongate body.

Family Salticidae—*Menemerus bivittatus* (Dufour)

Identifying Characteristics: This spider was previously known under the name *Marpissa melanognathus* (Comstock, 1912). *Menemerus bivittatus* is a dark colored spider with distinctively light markings. The cephalothorax is black to dark brown above with a median white mark posterior to the posterior eyes. In some males this white thoracic spot extends forward between the posterior eyes (Comstock, 1912). Each side of the cephalothorax is bordered with a thin white band. The abdomen is lighter than the cephalothorax, especially among males. A dark gray-rufus band extends along the midline of the abdomen and is bordered by dull white bands. A thin dark rufus band delineates the lateral margins of the abdomen. In the female the entire upper surface of the abdomen is covered with mixed gray and rufus hairs. Of taxonomic importance to the genus are four pairs of ventral spines on tibia I and a simple tooth on the retromargin of the cheliceral fang furrow.

FIGURE 202. A female *Menemerus bivittatus* displays her enlarged white colored palps.

Ecology and Behavior: *M. bivitattus* is commonly encountered among buildings, walls, barns, and abandoned construction sites. We have observed it during the daytime, out in the open, patrolling outside walls of stone buildings in search of prey. When disturbed, it retreated into the crevices in the walls. It is a southern species, and has been known to populate Florida (Comstock, 1912). The photographed specimen was collected just west of the Florida panhandle at the southernmost extension of Alabama.

Size: Length of female 6 to 8 mm; of male 5 to 7 mm.

Family Salticidae—*Metacyrba taeniola* (Hentz)

Identifying Characteristics: As the specific name (*taeniola)* infers, this is the "flattened" *Metacyrba*. Additionally, this spider is recognized by its

FIGURE 203. Lateral view of *Metacyrba taeniola* showing the black body with contrasting white stripes.

solid black cephalothorax. Some specimens are so black that discerning the black eyes is difficult. The abdomen is a charcoal gray with two narrow longitudinal, dull yellow lines. As in the photographed specimen, these lines may be broken into dots in some specimens. The femurs of the first pair of legs are flattened and somewhat stouter than the femurs of the other legs.

Ecology and Behavior: This spider is commonly found beneath loose tree bark and stones. However, they frequently take up residence in basements, garages and barns. The photographed female had for several months resided in my radial arm saw (RLJ). Whenever I would put my saw to use, she would emerge from hiding to contemplate my activities. Since her collection, I have noticed her produce numerous offspring throughout my property. *M. taeniola* can be collected throughout the eastern U.S. and as far west as California.

Size: Length of female 6 to 7 mm; of male 5 to 6 mm (Kaston, 1978).

Family Salticidae—*Pelegrina exigua* (Banks)

Identifying Characteristics: This species was previously known as *Metaphidippus exiguus.* The most distinctive features of this small jumping spider are the five pairs of black spots on the light brown abdomen. The most anterior pair of black spots is often smaller or less distinct. Interspersed between the black spots on each side of the abdomen are white spots. This species is distinguished from other closely related species in that the abdominal white spots do not cross the abdominal midline to form a transverse band. The cephalothorax is dark brown above and lighter brown along the sides. The legs are a dingy yellow which often have faint bands at the distal ends of the femur, patella and tibia. The chelicerae are the same yellow color as the legs and exhibit a black club-shaped mark when the spider is viewed face on.

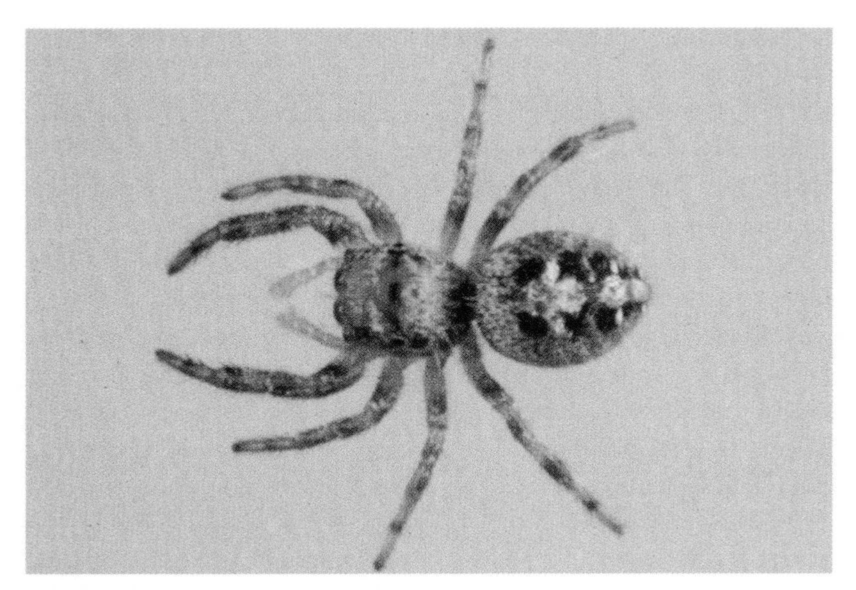

FIGURE 204. A female *Pelegrina exigua.*

Ecology and Behavior: These spiders are among the most common jumping spiders collected in sweep nettings of grass and brush from the eastern and southeastern U.S. We routinely collect this species from roadside patches of weeds throughout the summer and fall months and until the first winter frost. Females construct silken retreats within a folded leaf during the summer months. In the fall she deposits an egg sac inside her retreat. Young hatch and grow to a preadult size before winter. *P. exigua* ranges from New England and adjacent Canada south to Georgia and west to Ohio (Kaston, 1978). Jackman (1997; 1999) reported it from Texas.

Size: Length of female 4 to 5.6 mm; of male 3.3 to 5.1 mm (Kaston, 1978).

Family Salticidae—*Pelegrina galathea* (Walckenaer)—Peppered Jumper

Identifying Characteristics: The Peppered Jumper has a high and convex cephalothorax in which the sides bulge slightly in the region of the third row of eyes. The thirty-five known species of the genus have considerable variation in pigmentation and markings. Female *P. galathea* have a pair of dark spots at the posterior end of the abdomen. Immediately anterior to the dark spots is a pair of transverse white bands. These transverse white bands are the most posterior pair of four pairs of white abdominal markings. The first, second and third pairs of white markings are more circular or comma shaped. In the male the carapace is dark chestnut to black and the abdomen is reddish brown, and bordered in white. The legs in both sexes are short, stout and conspicuously ringed.

Ecology and Behavior: The spiders of this genus are frequently collected jumping spiders from sweep netting of tall grasses and bushes. *P. galathea* ranges from the eastern U.S. and adjacent Canada to the Rocky Mountains (Kaston, 1978). In this wide range, it is more common in the southern states than the northern.

Size: Length of female 3.6 mm to 5.4 mm; of male 2.7 mm to 4.4 mm (Kaston, 1978).

Family Salticidae—*Genus Phidippus*

This genus, with over 50 species, represents one of the largest groups of jumping spiders in the U.S. Its members are certainly the most frequently encountered jumping spiders. Due to their large size, robust bodies, and excessive hairiness, the phidippid spiders are the most easily recognized. Many are brightly colored with red, orange, or iridescent green coloration over the cephalothorax and abdomen. The phidippid jumping spiders have been claimed to make good house pets. They have voracious appetites for ants, termites and small flies. Levi (1990) reported

FIGURE 205 A-B. The tiny Peppered Jumper, *Pelegrina galathea*. Male (**A**) and female (**B**).

FIGURE 206. This *Phidippus* displays the tufts of bristles on the head typical of the genus.

that one phidippid spider ate more than forty fruit flies in succession. This would be more than twice the spider's body weight.

Family Salticidae—*Phidippus apacheanus* (Chamberlin and Gertsch)

Identifying Characteristics: The entire dorsum of this species is uniquely bright yellow to orange. The venter and legs are black to dark brown except the metatarsi and tarsi which are slightly banded. This species may be confused with *P. cardinalis,* which is sharply marked with a red carapace and abdomen.

Ecology and Behavior: This species is common in grassy and low herbaceous habitats. The photographed specimen was collected during the fall months from the leaf litter of a southern deciduous forest. The spider is a southern species ranging from Florida to California (Kaston, 1978).

FIGURE 207. A male *P. apacheanus* exhibiting its typical orange cephalothorax and abdomen.

Size: Length of female 11 to 12 mm; of male 9 to 11 mm (Kaston, 1978).

Family Salticidae—*Phidippus audax* (Hentz)—Bold Jumper

Identifying Characteristics: According to the criteria set down by Peckham (1909), *Phidippus audax* is distinctive in having an all black cephalothorax with many long white hairs. The abdomen is black without red markings. The abdomen has a series of white markings: an anterior white band, a large central triangular to somewhat circular spot, and two pairs of transverse posterior bands often broken at the dorsum midline. The front legs are especially robust and very hairy. From a facial view, the chelicerae of mature spiders boldly stand out as iridescent green or blue.

FIGURE 208 A-B. *Phidippus audax,* the Bold Jumper. (**A**) dorsal view, and (**B**) head-on view showing distinctive iridescent chelicerae.

Ecology and Behavior: During the warmest months of the year, these very common jumping spiders may be seen scurrying about stones, logs, and manmade structures, such as buildings and old barns. They are often seen searching outside building walls for prey. They usually occur near cracks in building walls where they immediately retreat if disturbed.

According to Jackman (1999) this spider is also common in cotton fields and many other crops where it feeds on boll weevils, adults and larvae of the bollworm, tarnished plant bugs, tobacco budworms and cotton leafhoppers, etc. This spider is an aggressive predator and has been observed to pursue and capture prey many times its own size. An immature spider will often construct a silken retreat, usually in a folded leaf, in which it overwinters. The spider occurs along the Atlantic coast states and adjacent Canada west to California (Kaston, 1978).

Size: Length of female 8 to 15 mm; of male 6 to 13 mm (Kaston, 1978).

Family Salticidae—*Phidippus clarus* (Keyserling)—The Red and Black Jumping Spider

Identifying Characteristics: This spider is often confused with other *Phidippus* spiders, especially those that have a considerable amount of red on the cephalothorax and abdomen. The female *P. clarus* has a red cephalothorax and abdomen; however; the abdomen is marked with an anterior basal white band, two longitudinal black stripes spotted with white,and oblique white bands along the sides. The male *P. clarus* is distinctive by having a black or dark purplish-black cephalothorax. The abdomen is red with a more prominent black band than in the female. The margins of the black band contain three pairs of white or red spots. The front legs of the male are longer and more powerful than the first legs of the female.

Ecology and Behavior: This spider is common in grassy habitats with broom straw and small broadleaf herbs. They may be collected in large numbers in late spring and summer by using a sweep net in tall grasses. Males and females typically pair for mating in July and August and lay their egg sac within a silken cocoon spun within a folded leaf. The female can be found inside the cocoon, often after the spiderlings hatch, while the male "stands guard" in the nearby vicinity. This spider is com-

FIGURE 209 A-B. *Phidippus clarus.* (**A**) A beautifully colored male; and (**B**) female.

monly found in all of the eastern states, and also throughout most of the continental United States and southern Canada (Kaston, 1978).

Size: Length of female 8 to 10 mm; of male 5 to 7 mm (Kaston, 1978).

Family Salticidae—*Phidippus otiosus* (Hentz)

Identifying Characteristics: This spider is distinctively colored relative to all other jumping spiders. It has four basic colors: black, white, orange and green. The legs and palps are ringed with black and white bands. The eyes and cephalothorax are largely black. The anterior portion of the abdomen is margined in white. The middorsal portion of the abdomen is mostly black, but is interrupted by some white hairs anteriorly, and by a yellow-orange "V" in the middle of the abdomen. The "V" has its apex pointing toward the head. Posteriorly and along the sides of

FIGURE 210. A handsome (but hairy) male of *Phidippus otiosus.*

the abdomen are three pairs of orange bands which extend upward toward the middorsum, but fail to meet at the midline. The chelicerae are iridescent to metallic green in color. The male is extremely hairy, especially on leg I.

Ecology and Behavior: This jumping spider is very active during daylight hours and seeks insect prey among small shrubs, bushes and grasses. It is not a particularly common species. Kaston (1978) gave its range as "Maryland south to Florida and across the Gulf States to Texas."

Size: Length of female 10 to 15 mm; of male 9 to 11 mm (Kaston, 1978).

Family Salticidae—*Phidippus whitmanii* (Peckham and Peckham)

Identifying Characteristics: *Phidippus whitmanii* is a large jumping spider that is conspicuously colored bright red. *P. whitmanii* is often confused

FIGURE 211. A young male *Phidippus whitmanii* dressed in its full regalia of bright red, white and black.

with *P. apacheanus,* which also has prominent reddish-orange markings on the cephalothorax and abdomen. However, *P. whitmanii* has the entire dorsum of the cephalothorax and the abdomen red in coloration, except for a distinct white band on the base and sides of the abdomen and two pairs of small white spots on the posterior abdomen. Additionally, the male, on its otherwise red cephalothorax, has a black hairless region extending from the first to the second row of eyes. In the female, a light band extends from between the anterior eyes and nearly crosses the black hairless region (Comstock, 1940).

Ecology and Behavior: This spider is found in grassy and low herbaceous habitats. The photographed spider was collected in early fall from a red brick building surrounded by heavy vegetation. The brick wall provided the spider with excellent camouflage while the dense vegetation provided it with a quick retreat from predators. This spider is predominantly a spider of the northern U.S. (Comstock, 1940). The spider photographed above is from Alabama.

Size: Length of female 9 mm; of male 8 mm.

Family Salticidae—*Platycryptus undatus* (De Geer)

Identifying Characteristics: For many years, this species has been known under the scientific name *Metacyrba undata,* but the generic name *Platycryptus* is now used by arachnologists (Hill, 1979). The spiders of the genus *Platycryptus* are characterized by flattened and elongated cephalothorax and abdomen. The cephalothorax and abdomen are nearly equal in height (Emerson, 1962). *Platycryptus undatus* is a large jumping spider with two distinctive features. The entire first row of eyes bears a single "eye brow" which is red-brown in the female and gray to white in the male. Secondly, the abdomen bears a single broad, dull white median stripe running its length. This median stripe bears irregular scalloped margins, bordered in black. Taxonomically important in this genus is the presence of only two pairs of robust spines on the underside of tibia I.

FIGURE 212. Female *Platycryptus undatus* hunting for its prey along a wooden fence post.

Additionally, the coxae of the first pair of legs lie very close together, closer than the width of the labium. Some authors emphasize the importance of the position of the second row of eyes that is equidistant between the first and third rows of eyes.

Ecology and Behavior: *P. undatus* has been collected beneath the bark of trees, on fences, and on the outsides of buildings. They are common in all eastern states during the warmest months. In southern states it overwinters and survives beyond the first year. The photographed specimen was collected in October. The spider had emerged from its retreat to hunt during a warm afternoon. This spider ranges from the eastern U.S. and adjacent Canada to Texas and Wisconsin (Kaston, 1978).

Size: Length of female 10 to 13 mm; of male 8.5 to 9.5 mm (Kaston, 1978).

Family Salticidae—*Plexippus paykulli* (Audouin)—Pantropic Jumper

Identifying Characteristics: This is a black to dark-brown spider that bears a median light band extending almost the entire length of the body. In the female, the medial white band is less distinct in the ocular region of the head. Also in the female, the medial white band is interrupted on the abdomen by a medial dark line on the anterior half and indistinct transverse dark chevrons and white markings on the posterior half. This band is widest on the posterior half of the abdomen. In the male, the cephalothorax and abdomen are also bordered laterally by white bands. Taxonomically important for this species is the presence of a shorter tibia and patella of the third leg than the fourth leg. Additionally, the ocular quadrangle is narrow posteriorly due to closely placed posterior eyes.

Ecology and Behavior: *Plexippus paykulli* is found in warmer parts of the world. In the United States it is restricted to southern localities, extending from Georgia south into Florida and west to Texas (Kaston,

FIGURE 213. Male of the Pantropic Jumper, *Plexippus paykulli.*

1978). This species is especially competitive in certain localities. Dauphin Island, Alabama has an abundant population. In the fall of 1997, *P. paykulli* had completely monopolized the old brick walls of Fort Gaines, excluding every other species of salticid spiders. During the fall of 2000, we found this spider exceptionally common around all-night gasoline stations in and around New Orleans, Louisiana.

Size: Length of female 10 to 12 mm; of male 9.5 mm (Kaston, 1978).

Family Salticidae—*Sarinda hentzi* (Banks)

Identifying Characteristics: In general appearance and behavior this spider strongly resembles a fast moving brown ant. A deep transverse depression in the cephalothorax posterior to the ocular quadrangle gives the impression of distinct head and thorax. In some specimens, this declivity is accentuated with a lighter band. Similar to antennae, the first pair of legs are elongate and are held in front of the head. The abdomen may be marked with two transverse lighter bands. The more posterior of these lighter bands may lie in a slight constriction at mid-length of the abdomen. Laterally this band is continued with white scales. In both sexes the tarsus and tibia of the palps are enlarged and armed with stiff bristles. The legs are longer than those of most jumping spiders and are ringed with the dark and light colors of the body. Legs I, II, and III have a prolateral, longitudinal dark line from femur to metatarsus.

Ecology and Behavior: *Sarinda hentzi* is collected with sweep nets in grasses and weeds in close proximity to streams or ponds. When collected, this spider moves much faster than ants and scurries for the shelter and the camouflage of leaves and litter. We suspect that this jumping spider climbs blades of grasses and weeds to scout out prey and/or mates. It ranges from New England to Florida and west to Texas and Kansas (Kaston, 1978).

Size: Length of adults about 5 to 7 mm in both sexes (Kaston, 1978).

FIGURE 214 A-B. Dorsal (**A**), and lateral (**B**) views of the ant-mimic, *Sarinda hentzi.*

Family Salticidae—*Thiodina puerpera* (Hentz)

Identifying Characteristics: The genus is identified by the presence of four bulbous setae on the tibiae of the first pair of legs. The width of the carapace is not over three-fourths its length. The body coloration of *Thiodina puerpera* is much like that of *Thiodina sylvana* except that, instead of the oval white spot of *T. sylvana,* there is a white band running from between the dorsal eyes down the thorax; and under the dorsal eyes, there is one white band instead of the three lines, and there is only one white line on each side coming up from the lower margin (Comstock, 1940).

Ecology and Behavior: *Thiodina puerpera,* like *T. sylvana,* is encountered during the warmest months in the southern states. Our specimens most often are collected from tall grasses using sweep nets. They may also be found in shrubs and bushes from open woods. The species occurs in the

FIGURE 215. Male of *Thiodina puerpera.*

Gulf coast states and ranges northward into Pennsylvania (Comstock, 1940).

Size: Length of female up to 11 mm; of male 6 mm.

Family Salticidae—*Thiodina sylvana* (Hentz)

Identifying Characteristics: The sexes of *Thiodina sylvana* are similarly shaped but the male is darker in color. The entire body of the female is yellow to white, darkest in the eye-region. The eyes boldly stand out because they are mounted on dark spots. The yellow abdomen in the female resembles the male in that it has two white lines and longitudinal rows of scattered dark spots. In the mature male the cephalothorax varies in shades of red-brown but it will usually have a white spot in the ocular quadrangle. The dorsum of the abdomen will also vary in shades of brown but it will always have two dorsal longitudinal white bands bordered by small dark spots. Spiders of this genus possess three unusual features for a jumping spider. Their third pair of legs is longer than their fourth pairs of legs. The ocular triangle is nearly square with parallel sides. Most distinctive is the four bulbous sensory setae on each tibia of the first pair of legs. These setae are difficult to see without a good microscope but they are definitive of the genus.

Ecology and Behavior: *Thiodina sylvana* is commonly found during the warmest months in the southern states. We have collected and photographed specimens most often from tall grasses, shrubs, and bushes from open woods. However, these spiders are as likely to be found residing in the non-native shrubbery of residential areas as in native flora of the rural areas. The species ranges from North Carolina south to Florida and west to California (Kaston, 1978).

Size: Length of female 8 to 10 mm; of male 7 to 9 mm (Kaston, 1978).

FIGURE 216 A-B. (**A**) Female; and (**B**) male of *Thiodina sylvana*.

Family Salticidae—*Zygoballus sexpunctatus* (Hentz)

Identifying Characteristics: This very small spider is a dark species with prominently elevated posterior eyes. The cephalothorax ranges from a copper-brown to black. Between the posterior eyes is a spot of white scales. The abdomen is black and has a basal white band and six spots, justifying the specific name. The most posterior pair of abdominal white spots is on the latter fourth of the abdomen. The other two pairs of abdominal spots are on the anterior half and arranged transversely so that one pair is lateral. The lateral spots are variable and may be lacking in some specimens.

Ecology and Behavior: We have collected this species by sweep netting roadside weeds and grasses during August and September. In these ecological settings, this spider feeds on tiny lepidopteran larvae and insects such as freshly hatched caterpillars and aphids. Its range extends from

FIGURE 217. Female *Zygoballus sexpunctatus.*

New Jersey south to Florida and west to Texas (Kaston, 1978). It is most commonly collected in the southern regions of its range.

Size: Length of female 3.5 to 4.5 mm; of male 3 to 3.5 mm (Kaston, 1978).

X
Glossary

ABDOMEN. This is the more posterior of the two body regions of a spider.

ALE. Anterior lateral eyes.

AME. Anterior median eyes.

ANAL TUBERCLE. Projection on the posterior end of the abdomen on which the anus opens.

ANNULATE. Rings of pigmentation around a body part such as a leg.

ANTERIOR. Toward the front end.

APOPHYSIS. A projection, larger than a spine, on the pedipalp or legs.

APPENDAGES. Structures that extend away from the main spider body, such as legs and palps.

ARACHNIDA. Class in the phylum Arthropoda that includes spiders, mites, ticks, harvestmen, scorpions, whipscorpions, windscorpions, and pseudoscorpions.

ARACHNOLOGY. The field of scientific investigation dealing with arachnids.

ARANEAE. Order of arachnids that consists of the spiders.

ARANEOMORPHAE. One of the two infraorders of spiders that contains most modern spider families, such as the orbweavers, wolf spiders, jumping spiders, etc.

AUTOTOMY. The willful breaking off of an appendage by a spider, usually in response to a leg being held by an enemy.

BOOK LUNGS. Sacs filled with air and provided with leaf-like folds that provide maximum surface area for exchanges of respiratory gases; found in most spiders.

BOSS. A smooth bump or protuberance on the side of the chelicerae near their base, in certain groups of spiders.

BRISTLE. An elongated, thin extension of the cuticle which is more slender than a spine.

CALAMISTRUM. A series of curved, comblike bristles on the dorsal surface or retrolateral margins of metatarsus IV of cribellate spiders, used to comb silk from the cribellum.

CARAPACE. A shieldlike dorsal covering of the cephalothorax.

CEPHALOTHORAX. The more anterior of the two major divisions of a spider's body.

CERVICAL GROOVE. A transverse groove on the carapace which separates the head and thorax.

CHELICERA (pl. chelicerae). The first pair of head appendages consisting of a stout basal portion and a distal fang; considered to be the "jaws" of a spider.

CHELICERAL TEETH. Large to small projections on the pro- and retromargin of the chelicerae; smaller ones often referred to as denticles.

CLAW. This is a strong, curved, usually black structure found on the distal end of the tarsus. Usually two equal-sized claws are present on most spiders, with many families having a third, smaller claw located between and below these two.

CLAW TUFTS. Two clusters of hairs arising from distinct pad-like regions near the outer base of the paired claws; present in certain groups of spiders, e.g., clubionids.

CLYPEUS. The region between the anterior eye row and the anterior edge of the carapace.

COLULUS. A small, usually conical structure found anterior to and between the anterior spinnerets in certain groups of spiders, e.g., sicariids. It represents the evolutionary remnant of the cribellum.

CONDYLE. A swollen, raised area (boss) at the base of the chelicera; found in certain spider groups with stout chelicerae, e.g., Araneidae, Lycosidae.

COXA (pl. coxae). The basal-most segment of the leg or pedipalp which attaches the appendage to the body.

CRIBELLATE. Adjective referring to a spider that possesses a cribellum.

CRIBELLUM. A flattened, perforated, plate-like structure found at the base of the anterior lateral spinnerets, through which a specialized kind of silk (cribellate silk) is spun; found in certain spider groups, e.g., Amaurobiidae.

CYMBIUM. This represents the highly modified tarsus of the pedipalp which has been enlarged and hollowed out to contain the copulatory organs of the male spider.

DENTICLE. A small, toothlike structure usually found on chelicerae, legs or palps.

DISTAL. Situated away or far from the point of attachment; opposite of proximal.

DORSAL. Toward the back as opposed to the ventral or belly side.

DORSAL FURROW. A midline groove, depression or pigmented line behind the cervical groove on the carapace.

DORSAL GROOVE. Same as dorsal furrow (see above).

DORSUM. Pertaining to the back, or upper surface.

ECRIBELLATE. Term referring to a spider which lacks a cribellum.

EGG SAC. The silk covering which the female spider spins around her egg mass. The shape of the egg sac and the method of its attachment to the substrate are often characteristic of a species.

EMBOLUS. A terminal tube-like structure of the male copulatory organ through which spermatozoa pass into the seminal receptacle of the female.

ENVENOMATION. The act of a spider injecting venom into its prey.

ENDITES. A platelike extension of each coxa of the pedipalps, forming two opposing structures (the “maxillae”). These structures lie lateral to the labium (lip) and ventral to the mouth opening, and they oppose the chelicerae while chewing.

EPIGASTRIC FURROW. On the ventral side of the abdomen, a transverse groove which separates the region of the book lungs from the more posterior portion of the abdomen.

EPIGYNUM (pl. epigyna). A hardened plate-like structure found in the midline of the abdomen immediately anterior to the epigastric furrow and which is associated with the female reproductive openings. It forms the external female genitalia. Epigynal structures form important distinguishing taxonomic features for some spider groups.

FANG. The hollow, claw-like structure forming the distal segment of the chelicera, used for piercing prey and for injecting venom.

FEMUR (pl. femora). The third segment from the proximal end of the spider's leg or pedipalp.

FOLIUM. A pigmented area forming a distinctive pattern or design on the dorsum of the abdomen.

GENICULATE. An adjective describing the condition whereby the base of the chelicerae is bent at a right angle in certain kinds of spiders.

GENITALIA. Term usually referring to the external reproductive organs such as the enlarged palpal organ of the male and the epigynum of the female.

GENUS (pl. genera). In the taxonomic heirarchy, this is the taxonomic category or subdivision just below the Family-level taxon.

HACKLED BAND. Term referring to the appearance of the silken threads spun by the cribellum and combed by the calamistrum; a unique arrangement of silken threads produced only by cribellate spiders.

HUB OF WEB. That region of an orb web where the radial threads converge and are fastened in the center to form a central meshwork or "hub" of the web.

LABIUM. A platelike structure below the mouth and between the endites; the lower lip whose inside surface forms the floor of the oral cavity.

LAMELLA. A broad, triangular, toothlike projection on the promargin of the cheliceral fang furrow of certain spider species.

LAMELLIFORM. Adjective meaning "flattened" and referring to the flattened hairs in claw tufts of certain spiders, e.g., anyphaenids.

LATERAL. Term referring to the right and left sides.

LATERIGRADE. A sideways form of locomotion characteristic of crab spiders. The term also refers to the unique way in which the legs are turned in certain spiders so that the leg surface that is normally dorsal is angled posteriorly.

LORUM (pl. lora). Plates on the dorsal surface of the pedicel.

LUNG SLITS. Openings into the book lungs, usually located along the epigastric furrow.

MASTIDION. A small tubercle or tooth-like elevation on the anterior surface of the chelicera in some spiders.

MAXILLA (pl. maxillae). The endite, or expanded basal segment of the pedipalp.

MEDIAN OCULAR AREA. The area on the dorsum of the head created by drawing an imaginary line around the median eyes of the anterior and posterior eye rows.

METATARSUS (pl. metatarsi). The next to the last segment of the leg; the sixth segment counting from the basal leg segment or coxa.

MYGALOMORPHAE. One of the two infraorders of spiders that contains most of the primitive spider families, such as the trapdoor spiders, purseweb spiders, etc.

OCULAR QUADRANGLE. The area on the head created by drawing an imaginary line around the anterior and posterior eye rows.

OCULAR TUBERCLE. An elevated bump or protuberance on the head, in some spiders, on which the eyes are located.

OPISTHOSOMA. The posterior body region of a spider, essentially equal to the abdomen.

PALP OR PALPUS. See pedipalp.

PATELLA (pl. patellae). The fourth segment of a leg or pedipalp counting from the proximal end, or coxa.

PEDICEL. The slender, stalklike structure connecting the abdomen with the cephalothorax.

PEDIPALPS. The second pair of appendages of the cephalothorax, behind the chelicerae but anterior to the legs. These structures are used as tactile organs. The palpal structure is simple in the female and resembles a leg, but in the male it is modified for sperm transfer.

PLE. Posterior lateral eyes.

PME. Posterior median eyes.

POISON GLANDS. The paired glands in the cephalothorax which produce the spider venom. They each have a duct which passes through each chelicera and opens by a pore near the tip of the cheliceral fang.

POSTERIOR. Toward the rear end.

PROCURVED. Usually refers to curvature of the eye rows whereby the lateral eyes project farther forward than the median eyes.

PROMARGIN. The front (anterior) margin of the cheliceral fang furrow; the margin anterior to the fang.

PROMARGINAL TEETH. A row of teeth on the anterior margin of the cheliceral fang furrow.

PROSOMA. The anterior body region of a spider; essentially equal to the cephalothorax.

PROXIMAL. Situated toward or near the point of attachment; opposite of distal.

RASTELLUM. Clusters of rake-like spines or teeth on the side of the chelicerae of trap-door spiders which aid them in digging their burrows.

REBORDERED. Having the edge or margin thickened, as in the rebordered labium in some spiders.

RECURVED. Usually refers to curvature of the eye rows whereby the median eyes project farther forward than the lateral eyes.

RETROMARGINAL TEETH. A row of teeth on the posterior margin of the cheliceral fang furrow.

SCAPE. A median process or appendage-like structure extending, usually posteriorly, from the epigynum.

SCOPULA. A dense brush of hairs along the median edge of the endite or the ventral side of the tarsus and metatarsus of some spiders.

SCUTUM (pl. scuta). A sclerotized (hardened) plate-like area on the abdomen of some spiders.

SETA (pl. setae). A hairlike moveable projection of the integument.

SEXUAL DIMORPHISM. The condition in which males of a species differ from females of the same species in details of size, structure, shape and color.

SIGILLUM (pl. sigilla). One of the paired indented bare spots on the sternum of some spiders, e.g., purseweb spiders.

SINUATE. Wavy in form or outline; serpentine.

SPERMATHECA. A sperm storage organ in females.

SPINNERET. One of six (rarely four or eight) fingerlike appendages at the end of the abdomen and modified for the extrusion of silk threads.

SPIRACLE. The external opening of the respiratory tubes (tracheae) located on the ventral side of the abdomen at some point between the anal tubercle and the epigastric furrow.

SPURIOUS CLAWS. Stout, serrated bristles at the end of the tarsus.

STABILIMENTUM. A heavy band of silk threads spun in the web, usually near the center, by certain orb weaving spiders, e.g. araneids.

STERNUM. A sclerotic plate forming the ventral surface of the cephalothorax behind the labium and between the bases (coxae) of the legs.

STRIDULATING ORGAN. A body region with numerous parallel ridges which are rubbed by an opposing structure on the pedipalp or abdomen in order to produce a sound; present in some spiders, e.g., males of *Steatoda borealis.*

SUSTENTACULUM. A thick, strong spine curved upward at its tip and projecting from the ventral surface of tarsus IV near the claws in certain spiders.

TARSUS (pl. tarsi). The most distal segment of the leg or pedipalp.

TERGITE. A sclerotized plate on the dorsal surface of the abdomen in some spiders.

THORAX. The region of the cephalothorax posterior to the cervical groove and bearing the legs; in spiders, it is fused with the head to form the prosoma or cephalothorax.

TIBIA (pl. tibiae). The fifth segment of the leg or pedipalp, counting from the proximal segment of the appendage (coxa).

TRACHEAE. Highly branched or simple respiratory tubes which open at the spiracle(s) and permeate the body of the spider, transporting respiratory gases to and from the tissues.

TRICHOBOTHRIUM (pl. trichobothria). A long, fine hair originating from a hemispherical socket and projecting outward at a right angle to the surface of the leg in some spiders.

TROCHANTER. The second segment of the leg or pedipalp, counting from the proximal segment of the appendage (coxa).

TRUNCATE. Having the end of a body part squared off, rather than coming to a point.

TUBERCLE. A low, rounded bump or process.

VENOM. The toxic substance secreted by the venom gland of the spider.

VENTER. The under surface of the body, away from the back.

VISCID SPIRAL. The sticky part of a spider web.

XI
Literature Cited

Anderson, J. F. 1974. Responses to starvation in the spiders *Lycosa lenta* (Hentz) and *Filistata hibernalis* (Hentz). *Ecology* 55:576–585.

Archer, A. F. 1940. The Argiopidae or orb-weaving spiders of Alabama. *Mus. Pap. Geol. Surv. Ala.* 14:1–77.

Archer, A. F. 1941a. Alabama spiders of the family Mimetidae. *Pap. Mich. Acad. Sci.* 27:183–193.

Archer, A. F. 1941b. Supplement to the Argiopidae of Alabama. *Mus. Pap. Geol. Surv. Ala* 18:1–47.

Askenmo, C., A. von Broemssen, J. Eckman and C. Jansson. 1977. Impact of some wintering birds on spider abundance in spruce. *Oikos* 28:90–94.

Beatty, J. A. 1970. The spider genus *Ariadna* in the Americas (Araneae: Dysderidae). *Bull. Mus. Comp. Zool.* 139(8):433–517.

Berman, J. D. and H. W. Levi. 1971. The orb weaver genus *Neoscona* in North America (Araneae: Araneidae). *Bull. Mus. Comp. Zool.*141: 465–500.

Brady, A. R. 1964. The lynx spiders of North America north of Mexico (Araneae: Oxyopidae). *Bull. Mus. Comp. Zool.* 131(13):432–518.

Brady, A. R. 1975. The lynx spider genus *Oxyopes* in Mexico and Central America (Araneae: Oxyopidae). *Psyche* 82(2):189–243.

Breene, R. G., D. A. Dean, G. B. Edwards, B. Hebert, H. W. Levi, G. Manning and L. Sorkin. 1995. *Common Names of Arachnids.* The American Tarantula Society, 94 pp.

Breene, R. G., W. L. Sterling and D. A. Dean. 1988a. Spider and ant predators of the cotton fleahopper on wooly croton. *Southwest. Entomol.*13:177–183.

Breene, R. G., M. H. Sweet and J. K. Olson. 1988b. Spider predators of mosquito larvae. *J. Arachnol.* 16:275–277.

Bristowe, W. S. 1939,1941. *The Comity of Spiders.* 2 vols. London, The Ray Society. Republished 1968 by Johnson Reprint Corp., New York.

Bristowe, W. S. 1971. *The World of Spiders.* London, Collins.

Brown, K. M. 1981. Foraging ecology and niche partitioning in orb weaving spiders. *Oecologia* 50:380–385.

Carico, J. E. 1973a. The Nearctic spider genus *Pisaurina* (Pisauridae). *Psyche* (1972)79(4):295–310.

Carico, J. E. 1973b. The Nearctic species of the genus *Dolomedes* (Araneae: Pisauridae). *Bull. Mus. Comp. Zool.* 114(7):435–488.

Carico, J. E. 1976. The spider genus *Tinus* (Pisauridae). *Psyche* 83(1): 63–78.

Carico, J. E. 1993. Revision of the genus *Trechalea* Thorell (Araneae, Trechaleidae) with a review of the taxonomy of Trechaleidae and Pisauridae of the Western Hemisphere. *J. Arachnol.* 21(3):226–257.

Chamberlin, R. V. 1947. A summary of the known North American Amaurobiidae. *Bull. Univ. Utah* 38(8):1–31.

Chamberlin, R. V. and W. Ivie. 1944. Spiders of the Georgia region of North America. *Bull. Univ. Utah* 35(9):1–267.

Chamberlin, R. V. And W. Ivie. 1947. North American dictynid spiders: The *bennetti* group of *Amaurobius*. *Ann. Entomol. Soc. Am.* 40(1): 29–55.

Clarke, R. D. and P. R. Grant. 1968. An experimental study of the role of spiders as predators in a forest litter community. Part I. *Ecology* 49:1151–1154.

Comstock, J. H. 1912. The Spider Book. (Revised edition, 1940). Comstock Publ. Associates, Ithaca, New York, 729 pp.

Comstock, J. H. 1940. *The Spider Book.* (ed. W. J. Gertsch). Cornell University Press., Ithaca, New York, 729 pp., (also reprinted 1948 and 1975).

Coyle, F. A. 1971. Systematics and natural history of the mygalomorph spider genus *Antrodiaetus* and related genera (Araneae: Antrodiaetidae). *Bull. Mus. Comp. Zool.* 141:269–402.

Craig, C. L. and G. D. Bernard. 1990. Insect attraction to ultraviolet-reflecting web decorations. *Ecology* 71:616.

Dondale, C. D. and J. H. Redner. 1978. The crab spiders of Canada and Alaska (Araneae: Philodromidae and Thomisidae). In: *The Insects and Arachnids of Canada,* part 5, Can. Dept. Agr. Pub. No. 1663, 255 pp.

Dondale, C. D. and J. H. Redner. 1982. The sac spiders of Canada and Alaska (Araneae: Clubionidae and Anyphaenidae). In: *The Insects and Arachnids of Canada*, part 9. Can. Dept. Agr. Pub. 1724, 194 pp.

Dondale, C. D. and J. H. Redner. 1983. Revision of the wolf spiders of the genus *Arctosa* C. L. Koch in North and Central America (Araneae: Lycosidae). *J. Arachnol.* 11(1):1–30.

Dondale, C. and J. Redner. 1990. The wolf spiders, nursery-web spiders and lynx spiders of Canada and Alaska. In: *The Insects and Arachnids of Canada*, part 17. Can. Dept. Agr. Pub. 1856:1–383.

Edwards, G. B. and D. E. Hill. 1978. Representatives of the North American salticid fauna. *Peckhamia* 1(5):110–117.

Edwards, R. J. 1958. The spider subfamily Clubioninae of the United States, Canada and Alaska (Araneae: Clubionidae). *Bull. Mus. Comp. Zool.* 118:365–436.

Emerton, J. H. 1902. Common Spiders of the United States. Ginn and Co. Publ., The Athenaeum Press, Boston, 235 pp.

Fincke, O. M., L. Higgins and E. Rojas. 1990. Parasitism of *Nephila clavipes* (Araneae, Tetragnathidae) by an ichneumonid (Hymenoptera, Polyspinctini) in Panama. *Journal of Arachnology* 18:321–329.

Fink, L. S. 1984. Venom spitting by the green lynx spider, *Peucetia viridans* (Araneae, Oxyopidae). *J. Arachnol.* 12:372–373.

Fitch, H. S. 1963. Spiders of the University of Kansas natural history reservation and Rockefeller Experimental Tract. *Univ. of Kansas Mus. of Nat. Hist. Misc. Publ.* 33, Lawrence, KS, 202 pp.

Foelix, Rainer. *Biology of Spiders*. Oxford University Press. New York 1996. Pp 110–149.

Forster, R. R. and N. I. Platnick. 1985. A review of the austral family Orsolobidae (Arachnida, Araneae) with notes on the superfamily Dysderoidea. *Bull. Am. Mus. Nat. Hist.* 181(1):1–229.

Forster, R. R., N. I. Platnick and M. R. Gray. 1987. A review of the spider superfamilies Hypochiloidea and Austrochiloidea (Araneae, Araneomorphae). *Bull. Am. Mus. Nat. Hist.* 185:1–116.

Fox, I. 1937. The Nearctic spiders of the family Heteropodidae. *J. Washington Acad. Sci.* 27(11):461–474.

Gertsch, W. J. 1939. A revision of the typical crab spiders (Misumeninae) of America north of Mexico. *Bull. Am. Mus. Nat. Hist.* 76(7):277–442.

Gertsch, W. J. 1953. The spider genera *Xysticus, Coriarachne* and *Oxyptila* (Thomisidae: Misumeninae) in North America. *Bull. Am. Mus. Nat. Hist.* 102(4):417–482.

Gertsch, W. J. 1958. The spider family Hypochilidae. *Am. Mus. Novit.* 1912:1–28.

Gertsch, W. J. 1964. A review of the genus *Hypochilus* and a description of a new species from Colorado (Araneae: Hypochilidae). *Am. Mus. Novit.* No. 2203, 14 pp.

Gertsch, W. J. 1979. American Spiders. 2nd ed, Van Nostrand Reinhold Company, New York, 274 pp.

Gertsch, W. J. and F. Ennik. 1983. The spider genus *Loxosceles* in North America, Central America and the West Indies (Araneae, Loxoscelidae). *Bull. Am. Mus. Nat. Hist.* 175(3):263–360.

Gertsch, W. J. and S. Mulaik. 1940. The spiders of Texas. I. *Bull. Am. Mus. Nat. Hist.*, 77(6):307–340.

Gertsch, W. J. and N. I. Platnick. 1975. A revision of the trapdoor spider genus *Cyclocosmia. Am. Mus. Novitates* 2580:1–20.

Gertsch, W. J. and N. I. Platnick. 1980. A revision of the American spiders of the family Atypidae (Araneae, Mygalomorphae). *Amer. Mus. Novitates* 2704:1–39.

Gertsch, W. J. and S. E. Riechert. 1976. The spatial and temporal partitioning of a desert spider community, with descriptions of new species. *Amer. Mus. Novitates* 2604:1–25.

Greenstone, M. H. 1980. Contiguous allotopy of *Pardosa ramulos* and *Pardosa tuoba* (Araneae:Lycosidae) in the San Francisco Bay Region, and its implications for patterns of resource partitioning in the genus. *American Midland Naturalist* 104:305–311.

Griswold, C. E. 1987. A revision of the jumping spider genus *Habronattus* F. O. P.-Cambridge (Araneae; Salticidae), with phenetic and cladistic analyses. *Univ. Calif. Publs. Ent.* 107: 1–344.

Hill, D. E. 1979. The scales of salticid spiders. *Zool. Jour. Linnean Soc.* 65:193–218.

Hillyard, P. 1994. *The Book of the Spider: From Arachnophobia to the Love of Spiders.* Random House, Inc., New York, 218 pp.

Hodge, M. A. 1987. Macrohabitat selection by the orb weaving spider, *Micrathena gracilis. Psyche* 94:347–361.

Jackman, J. A. 1997. *A Field Guide to the Spiders & Scorpions of Texas.* Gulf Publishing Co., Houston, TX, 201 pp.

Jackman, J. A. 1999. *A Field Guide to the Spiders & Scorpions of Texas.* Lone Star Books, A Division of Gulf Publishing Co., Houston, TX, 202 pp.

Kaston, B. J. 1948. The Spiders of Connecticut. *State Geol. and Nat. Hist. Survey Bull.* 70:1–874. Updated edition 1981:1–1020.

Kaston, B. J. 1978. *How to Know the Spiders.* 3rd ed., Wm. C. Brown Co. Publishers, Dubuque, Iowa, 272 pp.

Leech, R. 1972. A revision of the nearctic Amaurobiidae (Arachnida: Araneidae). *Mem. Entomol. Soc. Can.* 84:1–182.

Lehtinen, P. T. 1967. Classification of the cribellate spiders and some allied families, with notes on the evolution of the suborder Araneomorpha. *Ann. Zool. Fenn.* 4:199–468.

Levi, H. W. 1967. Cosmopolitan and pantropical species of theridiid spiders (Araneae: Theridiidae). *Pacific Insects* 9(2):175–186.

Levi, H. W. 1973. Small orb-weavers of the genus *Araneus* north of Mexico (Araneae: Araneidae). *Bull. Mus. Comp. Zool.* 145(9): 473–552.

Levi, H. W. 1974. The orb-weaver genera *Araniella* and *Nuctenea* (Araneae: Araneidae). *Bull. Mus. Comp. Zool.* 146 (6): 291–316.

Levi, H. W. and L. R. Levi. 1962. The genera of the spider family Theridiidae. *Bull. Mus. Comp. Zool.* 127(1):1–71.

Levi, H. W., L. R. Levi, and H. S. Zim. 1990. *Spiders and Their Kin.* Golden Press, New York, 160 pp.

Louda, S. M. 1982. Inflorescence spiders: a cost/benefit analysis for the host plant, *Haplopappus venutus* Blake (Asteraceae). *Oecologia* 55:185–191.

Manley, G. V., J. W. Butcher and M. Zarik. 1976. DDT transfer and metabolism in a forest litter macro-arthropod food chain. *Pedobiologia* 16:81–98.

Martyniuk, J. 1983. Aspects of habitat choice and fitness in *Prolinyphia marginata* (Araneae: Linyphiidae): web-site selection, foraging dynamics, sperm competition and overwintering survival. Unpublished Ph.D. Thesis, State University of New York at Binghamton.

Mayr, E. and P. Ashlock. 1991. *Principles of Systematic Zoology*. 2nd ed., McGraw-Hill, Inc.

Millidge, A. 1980. The erigonine spiders of North America. Part I. Introduction and taxonomic background (Araneae: Linyphiidae). *J. Arachnol.* 8(2):97–107.

Milne, L. and M. Milne. 1980. *National Audubon Society: Field Guide to North American Insects* and Spiders. Alfred A. Knopf, New York (18th printing, 2000).

Moulder, B. 1992. *A Guide to the Common Spiders of Illinois*. Illinois State Museum, Springfield IL 62706, 125 pp.

Muma, M. H. and W. J. Gertsch. 1964. The spider family Uloboridae in North America north of Mexico. *Am. Mus. Novitates* 2096:1–43.

Nyffeler, M. & G. Benz. 1981. Okologishce Bedeutung der spinnen als Insektempradotoren in wiesen und Getreidefeldern. Mitteilungen der deutschen Gesellschaft fur allgemeine und angewandte. *Entomolgie* 3:33–35.

Nyffeler, M., D. A. Dean & W. L. Sterling. 1987a. Evaluation of the improtance of the striped lynx spider, *Oxyopes salticus* (Araneae: Oxyopidae), as a predator in Texas cotton fields. *Environ. Entomo.* 16:1114–1123.

Nyffeler, M., D. A. Dean & W. L. Sterling. 1987b. Predation by green lynx spider, *Peucetia viridans* (Araneae: Oxyopidae), inhabiting cotton and woolly croton plants in East Texas. *Environ. Entomo.* 16:355–359.

Nyffeler, M. D., D. A. Dean, and W. L. Sterling. 1989. Prey selection and predatory importance of orb-weaving spiders (Araneae: Araneidae, Uloboridae) in Texas cotton. *Environ. Entomol.* 18:373–380.

Opell, B. D. 1979. Revision of the genera and tropical American species of the spider Family Uloboridae. *Bull. Mus. Comp. Zool.* 148(10): 443–549.

Opell, B. D. 1983. Checklist of American Uloboridae (Arachnida: Araneae). *The Great Lakes Entomologist* 16(2):61–66.

Peakall, D. B. 1971. Conservation of web proteins in the spider *Araneus diadematus. J. Exp Zoology* 176 (1971): 257.

Peck, W. B. 1981. The Ctenidae of temperate zone North America. *Bull. Am. Mus. Nat. Hist.* 170(1):157–169.

Peckham, G. W. and E. G. Peckham. 1909. Revision of the Attidae of North America. *Trans. Wisconsin Acad. Sci.* 16:355–646.

Platnick, N. I. 1974. The spider family Anyphaenidae in America north of Mexico. *Bull. Mus. Comp. Zool.* 146(4):205–266.

Platnick, N. I. 1989. *Advances in Spider Taxonomy 1981–87:* a supplement to Brignoli's *A Catalogue of the Araneae Described Between 1940 and 1981.* Manchester Univ. Press, Manchester, England, 673 pp.

Platnick, N. I. 1993. *Advances in Spider Taxonomy 1982–1991 with Synonymies and Transfers 1940–1980.* Manchester Univ. Press, Manchester, England, 846 pp.

Platnick, N. I. 1997. *Advances in Spider Taxonomy 1992–1995 with Redescriptions 1940–1980.* New York Entomological Society (in association with American Museum of Natural History), New York, NY, 976 pp.

Platnick, N. I., J. A. Coddington, R. R. Forster and C. E. Griswold. 1991. Spinneret morphology and the phylogeny of haplogyne spiders (Araneae, Araneomorphae). *Am. Mus. Novitates* 3016:1–73.

Platnick, N. I. and C. D. Dondale. 1992. The ground spiders of Canada and Alaska (Araneae: Gnaphosidae). In: *The Insects and Arachnids of Canada. Part 19.* Canada. Dept. Agri. Publ. 1875:1–297.

Platnick, N. I. and M. U. Sadab. 1977. A revision of the spider genus *Herpyllus* and *Scotophaeus* (Araneae: Gnaphosidae) in North America. *Bull. Am. Mus. Nat. Hist.* 159(1):1–44.

Platnick, N. I. and M. U. Sadab. 1980. A revision of the spider genus *Cesonia* (Araneae: Gnaphosidae). *Bull. Am. Mus. Nat. Hist.* 165(4): 337–385.

Platnick, N. I. and M. U. Sadab. 1981. A revision of the spider genus *Sergiolus* (Araneae: Gnaphosidae). *Am. Mus. Novitates* 2717:1–41.

Platnick, N. I. and M. U. Sadab. 1982. A revision of the American spiders of the genus *Drassyllus. Bull. Am. Mus. Nat. Hist.* 173(1):1–97.

Platnick, N. I. and M. U. Sadab. 1988. A revision of the American spiders of the genus *Micaria* (Araneae, Gnaphosidae). *Am. Mus. Novitates* 2916:1–64.

Polis, G. A. and S. J. McCormick. 1986. Scorpions, spiders and solpugids: predations and competition among distantly related taxa. *Oecologia* 71:111–116.

Raven, R. J. 1985. The spider infraorder Mygalomorphae (Araneae): cladistics and systematics. *Bull. Am. Nat. Hist.* 182(1):1–180.

Rehnberg, B. G. 1987. Selection of spider prey by *Trypoxylon politum* (Say) (Hymemoptera: Sphecidae). *Canadian Entomologist* 119: 189–194.

Reiskind, J. 1969. The spider subfamily Castianeirinae of North and Central America (Araneae: Clubionidae). *Bull. Mus. Comp. Zool.* 138(5):163–325.

Richman, D. B. 1989. A revision of the genus *Hentzia* (Araneae, Salticidae). *J.Arachnol.*17(3):285–334.

Richman, D. B. and B. Cutler. 1978. List of the jumping spiders (Araneae: Salticidae) of the United States and Canada. *Peckhamia* 1(5): 82–109.

Riechert, S. E. 1974. The pattern of local web distribution in a desert spider: mechanisms and seasonal variation. *Journal of Animal Ecology* 43:733–746.

Riechert, S. E. 1982. Spider interaction strategies: communications vs. coercion. In: *Spider Communication: Mechanisms and Ecological Significance,* P. N. Witt and J. S. Rovner, eds., pp 281–315. Princeton University Press, Princeton, NJ.

Riechert, S.E. and L. Bishop. 1990. Prey control by an assemblage of generalist predators: spiders in garden test systems. *Ecology* 71:1441–1450.

Riechert, S. E. and J. Harp. 1987. Nutritional ecology of spiders. In: *Nutritional Ecology of Insects, Mites, and Spiders,* F. Slansky, Jr. & J. G. Rodriguez, eds., pp 645–672.

Riechert, S. E. and C. R. Tracy. 1975. Thermal balance and prey availability: bases for a model relating web-site characteristics to spider reproductive success. *Ecology* 56:265–285.

Roth, V. D. 1993. *Spider Genera of North America, with Keys to Families and Genera, and a Guide to Literature*, third edition. Am. Arachnol. Soc., Gainesville, FL, 203 pp.

Roth, V., D. Carroll and D. Buckle. 1986. Linyphiidae of North America North of Mexico: checklists, synonymy and literature cited. Privately printed, 62 pp. Available through the American Arachnological Society.

Rypstra, A. L. 1981. The effect of Kleptoparasitism on prey consumption and web relations in a Peruvian population of the spider *Nephila clavipes*. *Oikos* 37:179–182.

Rypstra, A. L. 1984. A relative measure of predation on web-spiders in temperate and tropical forests. *Oikos* 43:129–132.

Rypstra, A. L. 1986. Web spiders in temperate and tropical forests: relative abundance and environmental correlates. *American Midland Naturalist* 115: 42–51.

Sabath, L. E. 1969. Color change and life history observations of the spider *Gea heptagon* (Araneae: Araneidae). *Psyche* 76:367–374.

Schick, R. X. 1965. The crab spiders of California (Araneidae: Thomisidae). *Bull. Am. Mus. Nat. Hist.* 129(1):1–180.

Schoener, T. W. and C. A. Toft. 1983. Spider populations: extraordinarily high densities on islands without top predators. *Science* 219: 1353–1355.

Shear, W. A. 1970. The spider family Oecobiidae in North America, Mexico, and the West Indies. *Bull. Mus. Comp. Zool.* 140:129–164.

Spiller, D. A. and T. W. Schoener. 1988. An experimental study of the effects of lizards on web-site communities. *Ecological Monographs* 58:51–77.

Spiller, D. A. & T. W. Schoener. 1990. Lizards reduce food consumption by spiders: mechanisms and consequences. *Oecologia* 83:150–161.

Sunderland, K. D., N. E. Crook, D. L. Stacey and B. J. Fuller. 1987. A study of feeding by polyphagous predators on cereal aphids using ELISA and gut dissection. *Journal of Applied Ecology* 24:907–933.

Suwa, M. 1986. Space partitioning among the wolf spider *Pardosa amentata* species group in Hokkaido, Japan. *Researches on Population Ecology* 28:231–252.

Syrek, D. and B. Janusz. 1977. Spatial structure of populations of spiders *Trochosa terricola* Thorell, 1856, and *Pardosa pullata* (Clerck, 1758). *Ekologia Polska* 25:107–113.

Ubick, D. and V. D. Roth. 1973. Nearctic Gnaphosidae including species from adjacent Mexican states; index to synonymy and invalid names in Nearctic Gnaphosidae, including Mexico. *Am. Arachnol.* 9:1–21.

Uetz, G. W., A. D. Johnson and D. W. Schemske. 1978. Web placement, web structure, and prey capture in orb-weaving spiders. *Bulletin of the British Arachnological Society* 4:141–148.

Valerio, C. E. 1981. Spitting spiders (Araneae, Scytodidae, *Scytodes*) from Central America. *Bull. Am. Mus. Nat. Hist.*, 170(1):80–89.

Van Hook, R. I., Jr. 1971. Energy and nutrient dynamics of spider and orthopteran populations in a grassland ecosystem. *Ecological Monographs* 41:1–26.

Vollrath, F. 1988. Spider growth as an indicator of habitat quality. *Bulletin of the British Arachnological Society* 7:217–219.

Vollrath, F. 1993. Spinnenseide—Superwerkstoff der Natur. In: *Kuntstoffe im Automobilbau.* VDI-Verlag, Dusseldorf 1993:1–18.

Whitcomb, W. H., H. Exline, and R. C. Hunter. 1963. Spiders of the Arkansas cotton field. *Ann. Entomol. Soc. Am.* 56:653–660.

Wiley, E. O. 1981. Phylogenetics: the theory and practice of phylogenetic systematics. New York, John Wiley & Sons, Inc.

Wise, D. H. 1993. *Spiders in Ecological Webs.* Cambridge University Press, London.

Witt, P. N., M. B. Scarboro, D. B. Peakall and R. Gause. 1977. Spider web-building in outer space: evaluation of records from the Skylab spider experiment. *Amer. J. Arachnology.* 4 (1977): 115.

Young, O. P. and T. C. Lockley. 1985. The striped lynx spider, *Oxyopes salticus,* in agroecosystem. *Entomophaga* 30:329–346.

Index

A

B

C

D

E

F

G

H

K

L

M

T

U

V

W

X

Z

About the Authors

W. Mike Howell, Ph.D., is a professor of biology at Samford University in Birmingham, AL. He has authored over 50 scientific papers. His contributions to science include discovery of the first environmental androgens as pollutants, development of silver staining methods for study of human chromosomes, and development of a viewing tank for studying and photographing fish without harming them. He and wife Mary Kirkland have two children and three grandchildren.

Ronald L. Jenkins, Ph.D., is a professor of biology at Samford University in Birmingham, AL, where he has taught for 16 years. His interests and scientific publications span the field of biology including spider ecology, comparative biochemistry, ethnobotany and environmental endocrine disruptors. He is married to Kitty Noordermeer and they have two children, Ben and Anna-Lea.